Springer Tracts in Modern Physics 80

Ergebnisse der exakten Naturwissenschaften

W0225704

Springer Tracts in Modern Physics

Neutron Physics

Contributions by
L. Koester   A. Steyerl

With 40 Figures

Springer-Verlag Berlin Heidelberg GmbH 1977

Dr. Lothar Koester

Fachbereich Physik, Reaktorstation Garching, Technische Universität München,
Lichtenbergstraße, D-8046 Garching

Dr. Albert Steyerl

Fachbereich Physik, Technische Universität München,
James-Franck-Straße, D-8046 Garching

Manuscripts for publication should be addressed to:

Gerhard Höhler

Institut für Theoretische Kernphysik der Universität Karlsruhe
Postfach 6380, D-7500 Karlsruhe 1

*Proofs and all correspondence concerning papers in the process of publication
should be addressed to:*

Ernst A. Niekisch

Institut für Grenzflächenforschung und Vakuumphysik der Kernforschungsanlage Jülich
Postfach 1913, D-5170 Jülich

ISBN 978-3-662-15835-7   ISBN 978-3-540-37543-2 (eBook)
DOI 10.1007/978-3-540-37543-2

Library of Congress Cataloging in Publication Data. Koester, Lothar, 1922—. Neutron physics.
(Springer tracts in modern physics; 80). Includes bibliographies. 1. Neutrons—Scattering. 2. Scattering length.
(Nuclear physics). 3. Neutrons. I. Steyerl, A., 1938—. II. Title. III. Series. QC1.S797. vol. 80. [QC793.5.
N4628]. 539'.08s. [539.7'213]. 76-52461

© by Springer-Verlag Berlin Heidelberg 1977
Originally published by Springer-Verlag Berlin Heidelberg New York in 1977.
Softcover reprint of the hardcover 1st edition 1977

Contents

Neutron Scattering Lengths and Fundamental Neutron Interactions

By *L. Koester*. With 19 Figures

Very Low Energy Neutrons

By *A. Steyerl*. With 21 Figures

Neutron Scattering Lengths
and Fundamental Neutron Interactions

Lothar Koester

1. Introduction

In the last three decades neutrons, mainly slow neutrons with velocities ranging
between 100 and some 1000 m/s, have become an important and unique tool in both
nuclear and solid state physics. They are useful in the study of bound matter and
for the investigations of the interactions between neutrons and elementary particles,
such as the electron and the proton. In all cases, the main process is the scattering of slow neutrons, which will be described by various scattering cross sections
given in terms of a basic nuclear constant. This basic quantity is the neutron scattering length b. Beginning at the time at which thermal neutrons became available
from nuclear reactors in sufficient intensity, methods for measuring scattering
lengths were developed which are based on the methods developed earlier for X-ray
studies, or which are used in light optics. Thus was created a new research field,
the neutron optics. Like light optics, neutron optics is characterized by the same
fundamental phenomena, diffraction, refraction and interference effects. The basic
quantity for describing these effects is the refractive index. In neutron optics,
the refractive index is related to the neutron scattering length.

   Much of the research in the first period has dealt with the fundamental processes involved in neutron scattering such as the measurement of coherent scattering
lengths by interference effects. In this way and also by reflection of neutrons,
the scattering amplitudes for most of the elements have been determined. As the progress in the theories of the solid state and nuclear interactions required results
of higher accuracy, e.g. /1.1/, new methods for measuring coherent scattering amplitudes were developed.

   In this review I shall first give a brief outline of the basic quantities and
of small, but necessary corrections due to second-order effects (Sec. 2). Then I
shall deal with the experimental methods for the determination of scattering lengths
and cross sections, in particular with the new high-precision techniques (Sec. 3).
"New" in this respect has a relative meaning, because the "new" methods have their
roots and some examples in the oldest physical fields of mechanics and optics. Even
the most recent technique is based on a theory established 50 years ago. In Sec. 4
I shall report on the employment of the new instruments and methods for the

investigations of elementary neutron interactions. In addition, a compilation of the most recent values for the neutron-scattering parameters of the elements is given for the use in pure and applied neutron physics.

Thus, I hope to be able to demonstrate that the progress gained by the application of slow neutron for fundamental physics is based on the unique refinement of the experimental methods.

Modern elementary-particle physics, for instance, makes dramatic progress by applying higher and highest energies for the study of particle interactions. As it is well known, this way is very expensive, the experiments are very difficult and their interpretation is not always immediately reliable. In other words, the experimental results and their evaluation may involve uncertainties. The conclusions from the high-energy experiments, however, can be complemented or partly tested by comparison with results of zero-energy experiments for which the interaction parameters are well defined. Progress of the investigations in the zero-energy region, therefore, is no less important. This progress can be achieved by a drastic improvement of the accuracy for the investigations of neutron-particle interactions with slow neutrons. This is only one aspect. In addition, it may be of interest to have a look at that field of new knowledge which is opened up by attaining a higher level of experimental accuracy.

2. Nonmagnetic Scattering of Neutrons

2.1 Neutron Data

The neutron appears as a unique particle; it is distinguished by the absence of nearly all electrical properties, it possesses "no" electrical charge, "no" electrical dipole moment, and "no" electrical polarizability. To what extent "no" means "absolutely no", remains a fundamental problem, which has been and will be the object of many very refined experiments.

The manifestation of the free neutron is all mass with spin and magnetism. The mass experiences the action of gravity. In the following, we shall describe the state of the free neutron in the standard way by the mass m, the velocity v, and the wavelength λ, or the wave number, $k = 2\pi/\lambda = \lambda^{-1}$. The notation "slow neutrons" is used in this respect to characterize neutrons with energies (E) from 10^{-4} eV up to 1000 eV. In this region the following nonrelativistic relations are valid

$$\lambda(\text{Å}) = 3.9560 \cdot [v(\text{km/s})]^{-1} \quad \text{or} \quad k(\text{Å}^{-1}) = 1.588 \cdot [v(\text{km/s})]$$
$$\lambda(\text{Å}) = 0.2860 \cdot [E(\text{eV})]^{-1/2} \quad \text{or} \quad k(\text{Å}^{-1}) = 21.97 \cdot [E(\text{eV})]^{1/2}$$

The numerical constants are computed with values of the fundamental physical constants as compiled by COHEN and TAYLOR /2.1/.

2.2 Neutron Scattering by Single Particles

2.2.1 Scattering Amplitude, Fermi Scattering Length and Effective Range

A plane neutron wave when scattered by a single fixed nucleus (spin $I = 0$) appears in the range r from the scatterer with a relative amplitude $f(\theta,k)/r$ in the direction given by θ.

Fermi described the scattering process by inserting into the Schrödinger equation a point-like pseudopotential for a nucleus in a position defined by $\underline{R}$

$$V(\underline{r}) = 2\pi\hbar^2\mu^{-1}a\delta(\underline{r}-\underline{R}).$$

(2.1)

Then the solution in the Born approximation gives for slow neutrons $f(\theta) = -a$ (see, e.g. /2.2/). μ denotes the reduced mass of the neutron-nucleus compound and the constant a is called the scattering length, which can be a complex quantity. The real part may be either positive or negative depending on the energy of the incident neutron and the particular nucleus involved in the scattering. The imaginary part represents absorption. In most cases it is very small.

In a more rigorous treatment of the scattering process it has been shown that for slow neutrons with wavelength ≥ 1000 nuclear radii (0.1 Å) the relations between S-wave phase shift δ_0, scattering amplitude f_0 and scattering length are given by /2.3/

$$f_0 = (1/2ik)\left[\exp(-2i\delta_0)-1\right]$$

$$k\cot\delta_0 = -a^{-1}+k^2r_{eff}/s \text{ and hence}$$

$$f_0 = -a\left[1-i(ka)+(1/2)(ka)\cdot(kr_{eff})-(ka)^2+...\right].$$

(2.2)

The constant r_{eff} is a measure for the effective range of the interaction potential. The quantities a and r_{eff} are independent of the form of the potential, provided that a short range potential is assumed. The optical theorem for the cross section σ is satisfied in the case of pure elastic scattering too

$$\sigma = 4\pi k^{-1} \, Jm \, f_0 = 4\pi a^2\left[1-k^2a(a-r_{eff})+...\right].$$

(2.3)

2.2.2 Potential and Resonance Scattering

Taking into account resonance state scattering, the expression for the scattering amplitude must be written in the form of two terms

$$f = f_p + f_r = f_p + \Gamma_n/\left\{2kr\left[(E-E_r)-i\Gamma/2)\right]\right\}$$

(2.4)

f_p is connected with potential scattering and f_r with resonance elastic scattering. Γ_n and Γ denote the neutron and total width of a single resonance level, respectively, at neutron energy E_r (see, e.g. FESHBACH et al. /2.4/).

For slow neutrons the resonance parameters have magnitudes, such that in many cases $(E-E_r) \gg \Gamma_n$ and $\Gamma_n + \Gamma_\gamma = \Gamma \ll E_r$ are fulfilled. Γ_γ is the radiation width of the level. Then the scattering amplitude assumes the simpler form

$$f = R' + \Gamma_n(2k_r E_r)^{-1} \cdot E_r(E-E_r)^{-1} \tag{2.5}$$

wherein the constant R' denotes the potential scattering radius /2.4/. Experimental values of R' for numerous elements were given by SETH et al. /2.5/. With $E \to 0$ the scattering length becomes for the case of only one resonance level

$$a = R' - \Gamma_n/2k_r E_r = R' - a_r \tag{2.6}$$

where $a_r(fm) = 2276 \cdot \Gamma_n(eV) \cdot E^{-3/2}(eV)$ is the resonance scattering length. If many separated resonances exist, the resonance contributions must be summed.

For nuclei with nuclear spin $I \neq 0$, s-wave neutron capture leads to two compound states, with spins $I+1/2$ and $I-1/2$. Then a resonance scattering length is associated with each of these spin states according to $a_r = g_+ a_r^+ + g_- a_r^-$ where $g_+ = (I+1)/(2I+1)$ and $g_- = I/(2I+1)$ are the statistical weight factors for $I-1/2$ and $I+1/2$ states, respectively. There are indications that R' depends on the spin of the compound state /2.6, 7/. This might be the case of such nuclei for which R' differs appreciably from $R = 1.5 \cdot 10^{-13} A^{1/3}$cm.

2.2.3 Neutron Scattering by the Coulomb Field of Nuclei

Neutron-Atom Scattering Length

The neutron-nucleus scattering is obviously due to nuclear forces. But the interactions of neutrons with the electric field of the nuclear charge also give non-vanishing contributions to the nuclear scattering length. The corresponding interaction potential for neutrons in an electrostatic field consists of two terms, namely (see e.g. /2.8/)

$$V_{LS} = -\gamma\mu_N(mc)^{-1}\underline{\sigma} \cdot \underline{E} \times \underline{p} \quad \text{and} \quad V_F = i\gamma\mu_N(mc)^{-1}\underline{E} \cdot \underline{P}$$

whereby $\gamma = -1.91$, μ_N is the nuclear magneton, m, $(1/2)\underline{\sigma}$ and $\underline{p}$ are the mass, spin and momentum of the neutron, respectively. $\underline{E}$ is the electric field seen by the neutron. The first term (V_{LS}) is usually referred to as the spin-orbit interaction /2.9/ (Schwinger-scattering) while the second term is the neutron's analogue of the Darwin interaction for electrons. The existence of this term was first pointed out by FOLDY /2.10 /.

The scattering amplitude related to the spin orbit potential is a purely imaginary quantity

$$f_{LS}(\lambda,\theta) = i \left[(\underline{\sigma} \cdot \underline{n})(m_e/2m)(\gamma e^2/m_e c^2) \cdot Z \ 1 - f_e(\lambda^{-1} \sin\theta/2) \ \cot\theta/2 \right] \tag{2.7}$$

which is opposite in sign for the two orientations of the neutron spin $(1/2)\underline{\sigma}$ with respect to the scattering plane defined by the normal unit vector $\underline{n}$. m_e denotes the electron mass, $f_e(\lambda^{-1}\sin\theta/2)$ is the X-ray formfactor of the electron configuration around the nucleus by which here the screening of the Coulomb potential is taken into account.

In the limit case of zero neutron energy, f_{LS} becomes zero and, hence, it does not contribute to the value of the scattering length. However, for precision determinations of the nuclear cross section ($\sim f^2$) at energies in the keV region the contribution due to f_{LS}^2 must be considered /2.11/.

The scattering length for the interaction of neutrons with the purely electrostatic potential of a bound electronic charge $(-e)$ was calculated by Foldy /2.10/ to be

$$b_F = \gamma(m_e/2m)(e^2/m_e c^2) = -1.468 \cdot 10^{-3} \ fm \tag{2.8}$$

Thus the electrostatic field of a bound atom leads to contributions to the neutron-atom scattering length b due to the nuclear charge $(+Ze)$: $-Zb_F$ and due to the shell-electron distribution: $ZF(\lambda)b_F$. $F(\lambda)$ represents the integrated X-ray formfactor (f_e) of the atom according to

$$F(\lambda) = (1/4\pi) \int_0^{2\pi} d\rho \int_0^\pi d\theta \cdot \sin\theta \cdot f_e(\lambda^{-1}\sin\theta/2) \tag{2.9}$$

If, in addition, an intrinsic neutron-electron scattering length b_e exists the total scattering length for the electron must be written as

$$b_{ne} = b_F + b_e \tag{2.10}$$

Hence, the atomic scattering length is given by

$$b(\lambda) = b_{NF} - Zb_F + ZF(\lambda)b_{ne} \tag{2.11}$$

or

$$b(\lambda) = b_{NF} - Zb_F \left[1 - F(\lambda) \ b_{ne}/b_F \right] \tag{2.12}$$

where b_{NF} is referred to as nuclear force scattering length. The combination $b_N = b_{NF} - Zb_F$ is called the scattering length of the bound nucleus. It describes the scattering if the neutron wavelength is so small that $F(\lambda)$ vanishes.

2.2.4 Nuclear Force Interaction

Of interest for nuclear physics is obviously the nuclear force scattering length b_{NF} defined by (2.11). At long neutron wavelengths for which $F(\lambda)$ becomes unity, the value of b_{NF} equals the atomic scattering length b which, in principle, can be accurately measured. However, a small difference between b_{NF} and b occurs if b_{ne} departs from b_F, in other words, if there exists an intrinsic (n,e) interaction. As is later shown, this interaction could be established with a good accuracy in a recent work /2.12/. With the result $b_{n,e} = -1.38 \pm 0.02 \; 10^{-3}$ fm the expression for b_{NF} becomes

$$b_{NF} = b - 8.95 \cdot 10^{-5} \cdot Z \; fm$$

This quantity is related to the mean nuclear potential $<V>$ within the range R of nuclear forces by the expression (see /2.13/)

$$b_{NF} = (2/3)(m/\hbar^2)R^3<V> \qquad (2.13)$$

Although the value of b_{NF} is independent of the shape of the potential some information about the potential can be deduced from the magnitude and sign of the scattering length. For instance, if b>R the potential is attractive and strong enough to form a bound state with the neutron. For b<o the potential is attractive but not strong enough to form a bound state. In the case o<b<R the potential is repulsive.

2.2.5 Coherent and Incoherent Scattering

Interference between incoming and scattered neutron waves and between the scattered waves from a set of nuclei is possible if the scattering by an individual nucleus is elastic and coherent. Fluctuations of the individual phase shift of the scattered waves, however, produce an incoherent component of scattering. Consequently, the scattering length a is separated into a coherent and an incoherent part, namely

$$a_c = <a> \quad \text{for the coherent and} \quad a_{inc} = (<a^2>-<a>^2)^{1/2} \quad \text{for the incoherent part.}$$

The brackets < > represent averaging over the individual scattering lengths. Fluctuations of phase shifts can be caused by resonance scattering with different spin states and by scattering on an assemblage of various isotopes of a chemical element or of various chemical elements.

In the case of spin-state scattering we have

$$a = g_+a_+ + g_-a_- \quad \text{and} \quad a_{inc} = (g_+g_-)^{1/2}(a_+-a_-)$$

Averaging over the isotope - or element - distribution yields

$$a_c = \sum_{j=1}^{s} \rho_j a_j \quad \text{and} \quad a_{inc} = \left[\frac{1}{2} \sum_{j}^{s} \sum_{j'}^{s} \rho_j \rho_{j'}(a_j-a_{j'})^2\right]^{1/2}$$

where s is the number of isotopes; ρ_j is the proportion of the j-th isotope and a_j is the scattering amplitude for th j-th isotope averaged over the spin states of this nucleus and the neutron.

2.2.6 Recommended Notations for Nonmagnetic Scattering Parameters

In the preceding considerations we have defined·several expressions for the scattering parameters of slow neutrons. After the considerable increase of the accuracy with which scattering parameters can now be determined (see following sections), it has become essential to distinguish clearly between the defined quantities for free and bound atoms or nuclei. Otherwise a great confusion may result. In particular, the contributions due to the neutron-electron scattering must be considered since the absolute error of $\pm$ 0.001 fm of accurate scattering lengths is of the same order as b_{ne} for a single electron.

Usually one denotes the scattering length by the symbol "a" for free particles and by "b", $b = (1+m/A)a$, for bound particles. The scattering amplitude, in general depending on the neutron energy or wave length, is denoted by $f(\lambda)$ or by $a(\lambda)$ and $b(\lambda)$. The latter is sometimes more convenient.

Recommended notations and some basic relations between the quantities are compiled in Table 2.1.

2.3 Refractive Index

So far, we have considered the neutron interaction with individual particles, atoms or nuclei. In a system of many fixed scatterers (N per cm^3), the multiple scattering of neutrons leads to additional phenomena. Thus, in the case of elastic coherent scattering, the interaction between neutrons and a large system of scatterers is described by the introduction of the index of refraction. Neutrons with wavelengths large in comparison with the distance between scatteres "see" in the scattering medium a mean potential energy

$$<V> = 2\pi\hbar^2m^{-1}N\cdot b \qquad (2.14)$$

which follows immediately from (2.1). With the kinetic energy $E = \hbar^2/(2m\lambdabar^2)$ of the neutron the refractive index n is expressed in an elementary way by

$$n^2-1 = -\langle V \rangle/E = -4\pi\lambdabar^2 N \cdot b .$$
(2.15)

Table 2.1. Recommended notations for scattering parameters of slow neutrons (b, a in 10^{-13} cm = fm, E in eV, λ in Å)

Interaction	Symbol and relations
Electron charge (Foldy)	$b_F = -1.468 \cdot 10^{-3}$
Electron	$b_{n,e}$
Electron (intrinsic)	$b_e = b_{n,e} - b_F$
Atom	$b(\lambda)$
Nuclear force	$b_{NF} = b(\lambda)+Zb_F\left[1-F(\lambda)b_{n,e}/b_F\right]$
	$b_{NF} = b-8.95 \cdot 10^{-5} \cdot Z$
Nuclear potential	$a_p = R' = R(1-\alpha)$
Nuclear radius	$R = 1.5 \cdot Z^{1/3}$
Nuclear resonance scattering	$a_r^{\pm}(E) = 2276 \sum\limits_i r_{ni}^{\pm} E_{ri}^{-3/2} E_{ri}/(E-E_{ri})$
Spin state scattering	$a^{\pm} = R'+a_r^{\pm}$
Nucleus (spin I)	$a_N = g_+a^+ + g_-a^-$
Statistical weight	$g_+ = (I+1)/(2I+1); \quad g_- = I/(2I+1)$
Atom of	$b(\lambda) = b_N + ZF(\lambda)b_{n,e}$
Element (mass number A)	$= b_{NF} - Zb_F\left[1-F(\lambda)b_{n,e}/b_F\right]$
Spin incoherent	$b_{inc} = (1+m/A)(g_+g_-)^{1/2}(a^+-a^-)$
Isotopic incoherent	$b_{inc} = (1+m/A)\left[1/2 \sum\limits_j \sum\limits_{j'} \rho_j \rho_{j'} \cdot (a_j - a_{j'})^2\right]^{1/2}$
Coherent	$b = (1+m/A) \sum\limits_i (W_{i+}a^+ + W_{i-}a^-)$
	$W_{i+} = \rho_i g_+; \quad W_{i-} = \rho_i g_-$

By rigorously treating the problem of multiple scattering and radiation damping, EKSTEIN /2.14/ calculated for cubic crystals and approximately for liquids the expression

$$n^2 - 1 = -4\pi\lambda^2 N \, b(1+3 \, b/\beta)^{-1} \tag{2.16}$$

where β is the lattice spacing of the crystal or the distance $V^{1/3}$ between scatterers in a liquid with the atomic volume V. For bismuth the correction term corresponding to b amounts to $0.7 \cdot 10^{-3}$ fm which is of the same magnitude as the experimental error of b.

The cross section σ_a for inelastic and absorption processes of the single fixed scatterer contributes an imaginary part to the refraction index. Herewith, but without the Ekstein correction, the complete expression follows /2.15/

$$n^2 - 1 = -N(\lambda^2/\pi) \cdot \left\{ \pm \left[b^2 - (\sigma_a/2\lambda)^2 \right]^{1/2} + i\sigma_a/2\lambda \right\} \tag{2.17}$$

Considering practical cases, the contribution of absorption and inelastic scattering to the refractive index is almost always negligible, whereas the effect on reflection curves at total reflection can not be neglected.

2.4 Neutron Scattering Cross Sections

2.4.1 Free Cross Section

Considering again the scattering of a plane neutron wave by a nucleus we find the' intensity of the scattered wave to be proportional to the scattering amplitude squared. Consequently, the corresponding cross section for scattering into a solid angle element $d\Omega$ is defined by $d\sigma = f(\theta,k)^2 d\Omega$. For the isotropic (s-wave) scattering of slow neutrons the total cross section is then

$$\sigma(k) = \int |(k)|^2 d\Omega = 4\pi |f(k)|^2 \tag{2.18}$$

Thus a corresponding cross section can be defined for each scattering amplitude or length introduced in the preceding sections, but, in general, only the total cross section at a neutron energy E for a dense system of many atoms is directly measurable. For atoms with individual scattering lengths a_i the mean cross section (per free atom) is usually expressed in terms of a by (see e.g. /2.16/)

$$\sigma(E) = 4\pi <a>^2 S_{coh}(E) + 4\pi (<a^2> - <a>^2) S_{inc}(E)$$
$$= 4\pi <a^2> S_{inc} + 4\pi <a>^2 (S_{coh} - S_{inc}) \tag{2.19}$$

where the angular brackets mean averaging over all individual scattering lengths. The contributions of solid state effects due to arrangement and dynamic behaviour of the atoms (mass M) in the sample are taken into account by the total (elastic and inelastic) scattering functions S_{inc} and $\delta S = S_{coh}-S_{inc}$. The behaviour of these function depends on the neutron energy E.

In the eV-energy range, where the neutron energy E is greater than the binding energy of atoms in matter, S_{inc} becomes (nearly) unity and δS (nearly) vanishes. Then the total cross section is given approximately by

$$\sigma(E) = 4\pi(1+m/M)^{-2} \{b-Z\left[1-F(E)\right]b_{ne}\}^2 \tag{2.20}$$

This relation between σ, b and b_{ne} allows an absolute determination of the scattering lengths by measuring σ (E).

In the limiting case $F(E_f) = 0$ the expression for $\sigma(E_f)$ becomes simply that for the free cross section $\sigma_{free} = 4\pi a_N^2(E_f)$. Thus, in general, $a_N^2(E_f)$ includes contributions due to resonance (Section 2.2.2) and Schwinger scattering (Section 2.2.3) and an effective range contribution (Section 2.2.1). Taking account of these, one obtains the zero-energy cross section

$$\sigma_o = 4\pi a_N^2(0) \tag{2.21}$$

For very exact determinations of a_N, further small contributions caused by solid state effects must be considered.

2.4.2 Corrections Due to Solid State Effects

For a calculation of S_{coh} and S_{inc} in the expression in (2.19) for the total cross section PLACZEK /2.17, 18, 19/ treated the coherent scattering as if it were incoherent. He calculated the correction terms to the incoherent approximation with $\delta S \approx 0$ in the two limiting cases of vanishing /2.20/ and high energy /2.21/ of the incident neutrons. A thorough calculation of δS was performed by BINDER /2.16/. His results for the various scattering cross sections of bismuth as functions of the neutron energy are presented in Fig.2.1. At energies higher than 0.1 eV the inelastic incoherent scattering characterized by S_{inc}^{inel} is predominant. The coherent part S_{coh}^{elast} in this energy range can be described in the incoherent approximation as S_{inc}^{elast} + δS^{elast} where δS^{elast} is referred to as the interference correction. It was previously calculated in accordance with Binder by PLACZEK et al. /2.21/ to be for heavy nuclei

$$\delta S^{elast} = -I\lambda^2/8\pi v_o^{2/3} \tag{2.22}$$

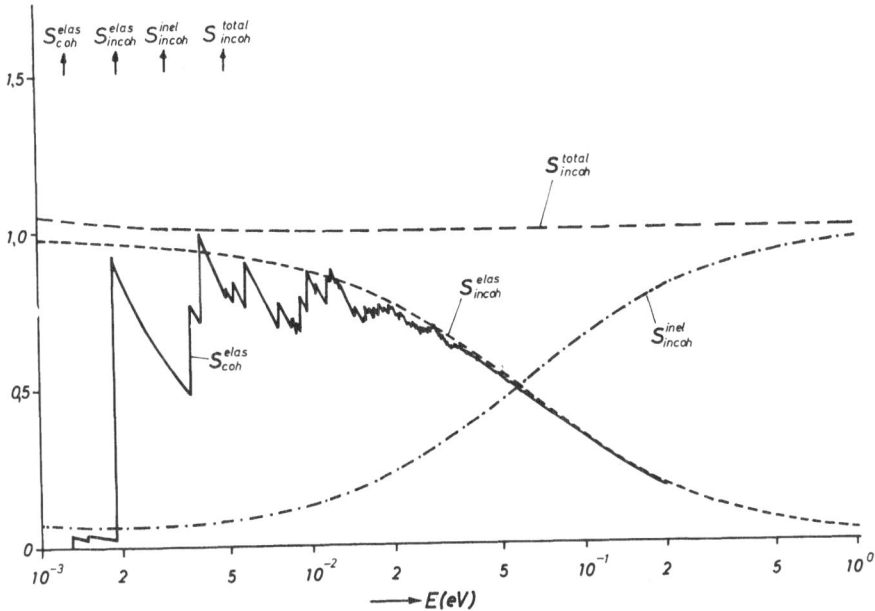

Fig. 2.1. Total cross section of bismuth as function of the neutron energy /2.16/

where v_o denotes the atomic volume. The numerical constant I depends on the lattice structure, typical values are I = 2.888 for cubic lattices and I ≈ 2.8 for a heavy liquid.

A further correction to σ_{free} is due to thermal motion of the atoms. The so-called Doppler correction gives a small contribution ΔS_{inc} to the incoherent nuclear cross section which was calculated by PLACZEK /2.17/ and WICK /2.22/ for systems of heavy and light nuclei, respectively. The results are given in terms of the free cross section by the expression

$$\sigma/\sigma_{free} = 1 + K_{AV} \ m/3M \cdot E -+ \text{smaller terms}, \tag{2.23}$$

where M denotes the atomic mass. K_{AV} is the mean kinetic energy of the atoms. Because of the smallness of the correction term one may take $K_{AV} \approx (3/2)k_B T$ (k_B = Boltzmann constant, T = temperature in K). Thus follows from (2.23) in a good approximation for higher energy (eV)

$$\sigma/\sigma_{free} \approx 1 + k_B Tm/2EM. \tag{2.24}$$

11

Both correction terms are correct for heavy nuclei rather than for light nuclei. But at higher neutron energies the magnitude of the corrections becomes so small that it may be neglected in most cases. In the most precise experiments, however, it must be considered.

In the other limiting case of very low energies ($\leq 10^{-3}$ eV), the cross section for isotropic scattering is mainly inelastic. For this case BINDER /2.16/ calculated the scattering functions with the Debye approximation (Debye temperature: Θ) and for one-phonon interaction, if the temperature T of the sample is not small compared to Θ, to be

$$S_{inc}^{inel} \cong (m/M)(k_B\Theta/E)^{1/2} \left[6T/5\Theta - 3/7 - \Theta/18T - (\Theta/T)^2/1560 - + .. \right] \qquad (2.25)$$

The interference correction term δS_{inc}^{inel} contains the phonon dispersion relation in a complicated manner, so that the Debye approximation can lead only to the order of magnitude of δS. For very slow neutrons and cubic crystals the following holds approximately: $\delta S_{inc}^{inel} \sim 0.1\ S_{inc}^{inel}$. Hence the inelastic cross section can be expressed for this case by

$$\sigma_{ie} \sim 4\pi <b^2> \cdot 1,1 \cdot S_{inc}^{inel}. \qquad (2.26)$$

Measurements of total cross sections at very low neutron energies and consequently the inelastic cross sections are of physical interest, as STEYERL has reported in the same volume. Moreover, in some cases such measurements may be suitable for determinations of small water contaminations in samples since the cross section per water molecule (about 270 barn at $5 \cdot 10^{-4}$ eV) is much greater than the inelastic cross sections which amount to 1 to 10 b per molecule of the sample /2.23/.

3. Methods for the Determination of Scattering Lengths

3.1 Bragg Diffraction Methods

So far, the majority of the data for coherent scattering lengths have been obtained from diffraction measurements under Bragg conditions. In this way the intensities of neutron beams scattered from single or powdered crystals have been measured using almost the corresponding methods developed for X-ray studies. In particular, the powdered-crystal method introduced for the case of neutrons by WOLLAN and SHULL /3.1/ has been most extensively used in diffraction work for the determination of scattering lengths (see, e.g. /3.2/).

The diffraction method and its extended use for the investigations of neutron scattering and of the structure of matter is intensively described in review articles /3.3, 4/ and books /3.5, 6/. We shall therefore make only a few general remarks con- concerning the accuracy of the determination of scattering length.

In a typical arrangement for powder-diffraction measurements, the powdered sample is in the form of a vertical circular cylinder, which is completely bathed in a mono- energetic primary neutron beam (wavelength λ). The number P of neutrons diffracted per minute under the glancing angle 2θ into the detecting counter is then given by

$$P_{hkl}/P_0 = \text{const}\cdot\lambda^3(V\rho'/\rho)jN_c^2F^2(\sin\theta\cdot\sin2\theta)^{-1}A_{hkl}\ \exp\left[-2W(T)\right] \qquad (3.1)$$

where P_0 = number of neutrons hitting the specimen per minute. ρ', ρ: measured and theoretical density of the specimen, j = number of cooperating planes for the partic- ular reflexion being measured, N_c = number of unit cells per cm^3, F = structure am- plitude factor per unit cell, $e^{-2W(T)}$ = Debye-Waller temperature factor for the reflection in question and A_{hkl} = absorption factor. The constant is determined by the geometry of the sample-detector arrangement.

The structure amplitude factor F contains the scattering lengths b_j of the m atoms in the unit cells as follows

$$F^2 = \sum_{j=1}^{m} \left\{(b_N+Zf_{hkl}b_{ne})_j^2 + (1-f_{hkl})^2\gamma^2ctg^2\theta_{hkl}\right\} \times$$
$$\exp\left[2\pi i(hx_j + ky_j + lz_j)\right]. \qquad (3.2)$$

The term containing $\gamma^2ctg^2\theta_{hkl}$ takes account of Schwinger scattering (Section 2.2.3). f_{hkl} is the usual X-ray formfactor. For the determination of F^2 according to (3.1.1) measurements of the integrated intensities P_{hkl} of Bragg reflexes as well as the measurement of the intensity P_0 hitting the specimen are necessary. Since all the other quantities in the expression can be measured too, it follows that the value of F^2 can be deduced on an absolute scale. The accuracy, however, is reduced by the experimental difficulties in measuring very accurately the neutron intensities, the geometrical constant and the absorption factor. In practice, it is easier to deter- mine the instrumental constants by making measurement on reference substances of known scattering amplitude factor.

In doing so, the results for other substances can be given only in terms of the reference value. In their pioneer work SHULL and WOLLAN /3.2/ gave an accuracy of 3 or 4 precent for the scattering length. Within the scope of the instrument opera- tion it is possible to attain at least an accuracy of 1 %. However, special pre- cautions as to sample purity and instrument performance must be taken.

3.2 Pendellösung Fringes

There are two general theories which may be used to account for the intensities observed in diffraction studies. The above-used kinematical theory treats the scattering from each volume element in the sample as being independent of the scattering
of other volume elements except for incoherent power losses in reaching and leaving
that particular volume element. The other theory, called the dynamical theory, takes
into account all wave interactions within the crystalline particle, and must generally be used whenever diffraction from large perfect crystals is being considered
(see e.g. /3.7/).

For the case that monochromatic neutron radiation is brought into a perfect
crystal under near-Bragg orientation, the dynamical theory predicts a coherent
splitting of the incident neutron wave within the crystal into four components,
two components travelling in the Bragg direction and two components in the forward
or incident direction. In general, each of the four components is to be described
by different wave vectors, differing in magnitude and direction as shown in Fig.3.1.

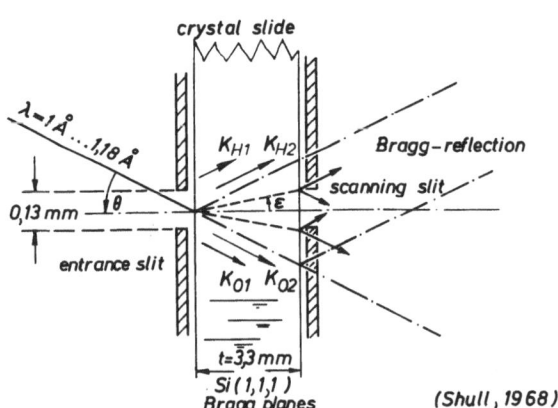

Fig.3.1. Diagram of ex-
perimental arrangement for
studying Pendellösung fringe
structure within the Bragg
reflection from a Laue
transmitting crystal /3.9/

The beams in incident and diffracted direction are coherently coupled and the energy
is swapped back and forth between them so that the total field must be considered as
a unit (Pendellösung structure). Among the constants which would determine the case
of transfer of radiation travelling one way to radiation travelling, another way would
be the scattering length b of the scattering centres present in the crystal. The
dynamical theory predicts that radiation energy is transmitted in two symmetrical

directions described by the angle ε which equals the difference between the direction of the incident ray θ and the exact Bragg angle θ_B. If the incident beams contain ray components encompassing all of the Bragg cone, the Bragg-reflected intensity emanates from the crystal surface over a certain band. The linear width of this band is determined by ε and the crystal thickness t_0. KATO /3.8/ and SHULL /3.9/ have shown that the intensity distribution within this band is a periodic function of $\tan\theta$ given by

$$I(\gamma) = C \cdot (1-\gamma^2)^{-1/2} \sin^2\left[(\pi/4)+A(1-\gamma^2)^{-1/2}\tan\theta\right] \qquad (3.3)$$

with $\gamma = \tan\varepsilon/\tan\theta_B$ and C = normalization constant. The quantity A, called Pendellösung period, is expressed as

$$A = 2t_0 N_c d(F_H\,F_{\bar{H}})^{1/2}\exp(-2W) \qquad (3.4)$$

where N_c is the number of unit cells per unit volume, $F_H\exp(-2W)$ is the crystal structure amplitude per unit cell and d the interplanar spacing.

When the neutron ray comes into the crystal at the exact Bragg angle ($\varepsilon = 0$ and $\gamma = 0$), the corresponding wave components in the forward and Bragg direction have the same magnitude. Thus they can form by superposition two parallel resultant waves whose wave vectors are exactly along the diffracting planes. The resulting energy propagation is straight through the crystal. At the exit surface the energy then splits into its two coherent components - one travelling in the Bragg direction, the other travelling in the forward direction. We will later see how this coherent beam splitting and the coherent superposition of individual beams within the crystal can be used for interference experiments with neutrons. In the exact Bragg case the intensity /3.3/ has the simpler form

$$I(\lambda) = C \cdot \sin^2\left\{(\pi/4) + \lambda t_0 N_c \cdot F_H\left[\exp(-2W)\right]/\cos\theta_B\right\} \qquad (3.5)$$

depending on the wavelength λ.

By varying λ under Bragg conditions the period of the fringe function $I(\lambda)$, which contains the structure amplitude F_H, can be determined. Since the method depends fundamentally on geometrical measurements of length and angle, the ultimate precision is limited only by the care in determining these quantities and by source intensity.

This diffraction method for measuring neutron scattering amplitudes was developed by SHULL /3.10/ who also investigated ingeniously and intensively the phenomenon of neutron dynamical scattering on perfect crystals /3.11/. Shull has attained accuracies of $\pm 3 \cdot 10^{-4}$ and $5 \cdot 10^{-4}$ $\Delta b/b$ for the scattering lengths of Si and Ge, respectively.

One should emphasize, however, that the experimental difficulties are extremely high and that this method is restricted to the few elements available in the form of perfect crystals.

3.3 Prism Refraction

As in light optics a neutron beam is deflected when passing through a prism /3.12/. Since the index of refraction is usually smaller than unity for neutrons, the bending of the beam is towards the apex angle (2α).
The magnitude of this angular deflection is given by

$$\beta = 2(1-n)\tan\alpha$$

and according to the expression for n (2.15) by

$$\pi\beta = Nb\lambda^2\tan\alpha. \tag{3.6}$$

In principle, this relation can be used to determine the scattering length b of the prism material by measuring the deflection angle β for neutrons of known wavelength. However, the deflection is very small. For instance, in the case of a prism $(2\alpha = 120^\circ)$ of copper, the deflection angle is calculated to be about 3 s arc for neutron of 2Å wavelength. Such small angles can be measured in the double crystal spectrometer first described by SHULL et al. /3.13/.

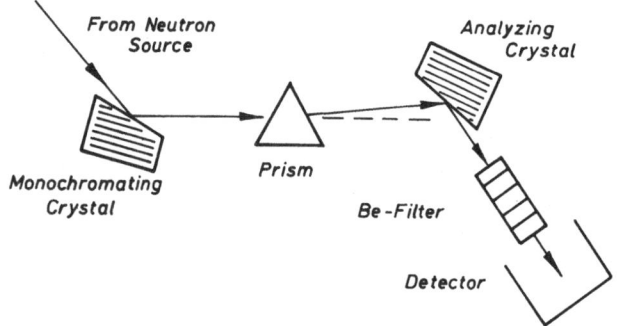

Fig.3.2. Schematic diagram of the double-crystal spectrometer used for measurement of prism refractive bending /3.14/

The principle of this spectrometer is shown in Fig.3.2 in which an arrangement with very high angular resolution is presented /3.14/. The apparatus consists of two perfect silicon crystals whose surfaces were cut 11° away from the (111) planes.

The incoming well-collimated neutron beam hits the monochromator crystal from which a Bragg-reflected beam emanates. After a certain range this beam strikes the analyzing crystal, which is in antiparallel orientation to the monochromazing crystal. The Bragg-reflected ray then passes on the way to the detector through a beryllium filter by which the shorter wavelengths neutrons from second- and higher-order reflections were removed from the beam. Because of the perfect structure and the antisymmetrical cut of the crystals, a very high angular collimation (up to 1 mm width per 1000 m range) of the reflected beam is obtained. By turning the analyzer crystal and observing the intensity with the detector, one measures an extremely sharp rocking curve. At neutron wavelength 2.393(3) Å a rocking curve with a full width at half maximum intensity of 1.32 s of arc was obtained. SCHNEIDER and SHULL /3.14/ used this spectrometer to measure the bending of neutrons by prism of unmagnetized and magnetized iron. They have reported the nuclear and the magnetic scattering lengths for the prism material with an accuracy in $\Delta b/b$ of ± 1 %. Most recently SCHNEIDER /3.16/ has reported on experiments on prisms of SiO_2, Cu, and Ge in which the scattering lengths could be determined with uncertainties of only 2 to 9 parts in 10^4.

In Fig.3.3 are shown, for example, rocking curves of an undeflected beam and of a beam deflected by a bi-prism of Cu as measured by Shull with his torsion-bar spectrometer /3.15/.

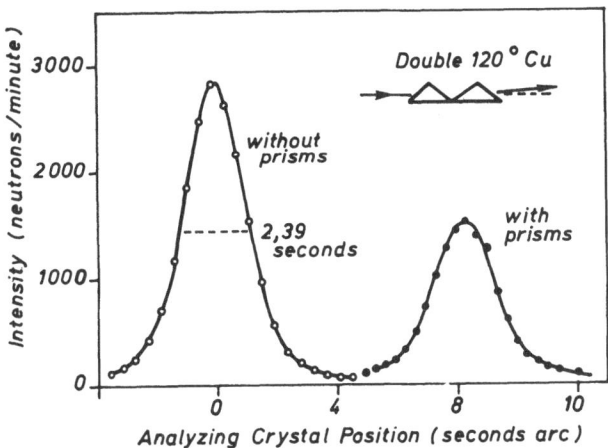

Fig.3.3. Rocking curves of the double-crystal spectrometer used for measurements of prism refractive bending /3.14/

3.4 Bi-prism Interferometer

MAIER-LEIBNITZ and SPRINGER /3.13/ reported for the first time on a Fresnel-type
interferometer for neutron radiation in which two coherent partial neutron beams
were obtained by refraction of a neutron beam by a bi-prism. The principle of this
interferometer in which the first interference pattern for neutrons was measured
is shown in Fig.3.4. A typical interference curve obtained by LANDKAMMER /3.16/ is
given in Fig.3.5. The instrument was thought to be suitable for very precise determinations of refractive indices for elements and isotopes. But all attempts to attain
interference patterns with different probes positioned in the coherent beams as shown
in Fig.3.4 have been unsuccessful. Even with two probes of the same material, interference effects could not be observed /3.16/. Recent experiments /3.17/ have shown
that the spatial distance of about 60 μm between the interfering beams is too small
for an exact adjustment of the interface of divided probes between the beams.

3.5 Perfect Crystal Interferometer

The main disadvantage of the bi-prism interferometer, namely, the extremely small
separation distance between the coherent partial beams, can be avoided by making use
of perfect crystals for the coherent beam splitting (Section 3.2) as well as for the
subsequent superposition of the individual beams. This principle was realized at
first in X-ray interferometers /3.18/, which have been used successfully over the
last decade. Interferometer application to neutron radiation has been proposed by
BONSE and HART /3.19/ RAUCH /3.20/ and SHULL /3.21/. The first perfect-crystal neutron interferometer was described by RAUCH et al. /3.22/. The principle arrangement
of their instrument is shown in Fig.3.6.

It consists of an E-shaped perfect crystal of silicon which was cut to give symmetrical (220)-Laue reflections in the three connected slabs. The emerging neutron
beam from a reactor is monochromized by Bragg reflection on a graphite crystal. The
beam strikes the first crystal slab and splits coherently in partial beams in the
forward and deflection direction. Both beams pass through the second slab and are
deflected again in the described manner, thus forming the beams I and II. In the
third slab an overlap of the convergent partial beams and Bragg reflection takes
place forming forward and deviated beams (0 and H). The crystal arrangement has a
length of 7 cm. The distance between the partial beams amounts to 4 cm. Interference
can occur in the overlap region if every atomic plane in the silicon crystal is on
the average in the right position within 10^{-8} cm.

Any material with thickness D (Fig.3.6) and refractive index $n = 1-\lambda^2 \cdot Nb/2\pi$ in
one coherent beam causes a phase shift of $(1-n)2\pi D/\lambda$. This leads to an intensity modulation for the (forward) 0-beam according to

$$I_0 \cos^2|\pi(1-n)D/\lambda|. \qquad\qquad (3.7)$$

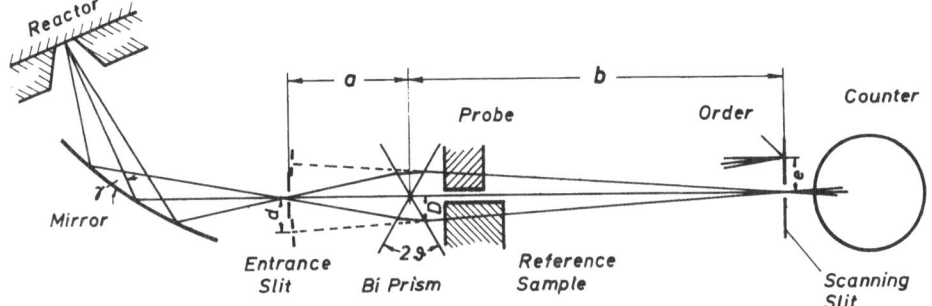

Fig.3.4. Schematic diagram of the neutron interferometer with Fresnel's bi-prism. Reflection angle $\gamma \simeq 0.1^{\circ}$, distance entrance to main slit: a + b = 10 m, maximum distance between the beams: 2 D = 0.1 mm /3.12/

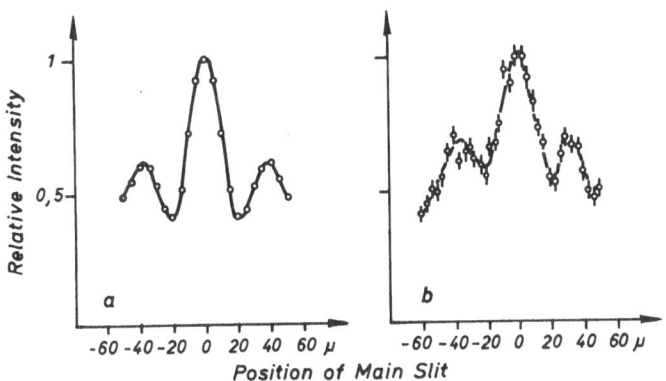

Fig.3.5 a and b. Interference pattern of the neutron interferometer. (a) calculated curve, (b) experimental curve /3.16/

The wavelength spread of the beam and all other inaccuracies of the device and of the phase shifting material result in a smoothing of the intensity modulation particularly for high order interferences (Fig.3.7). The expression (3.7) for the intensity modulations exhibits the experimental way for a determination of the refractive index of the phase shifting material. A continuous variation of the thickness D must result in a periodic variation of the neutron intensity in the 0-beam with the period

$$D_0 = \lambda/(1-n) = \frac{2\pi}{\lambda Nb} \; . \tag{3.8}$$

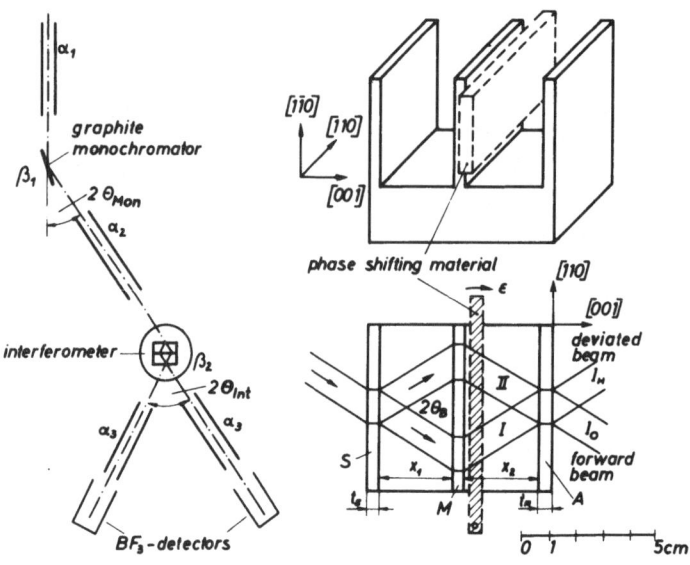

Fig.3.6. Sketch of the single crystal neutron interferometer. Left: total arrangement. Right: interferometer crystal /3.23/

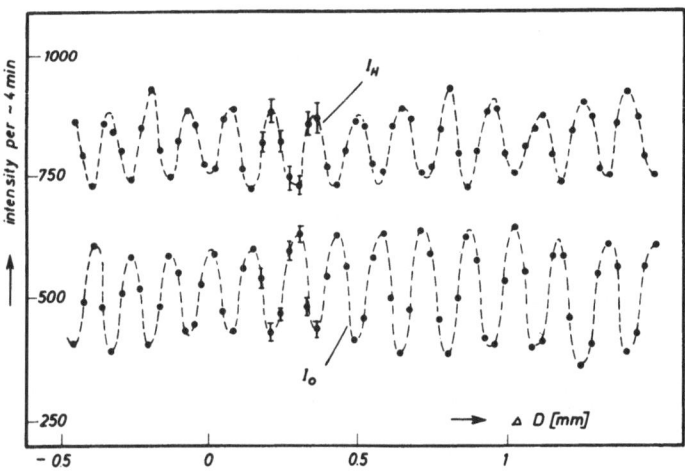

Fig.3.7. Observed intensity oscillation of the deviated (I_H) and forward (I_0) beam for Al as phase shifting material in the perfect-crystal interferometer /3.23/

In proper cases the value of this period can be measured with an accuracy in the order $\pm 5 \cdot 10^{-4}$ $\Delta D_0/D_0$. To obtain a corresponding accuracy for the scattering length b the wavelength λ and the atomic density N (atoms per cm^3) must be determined with similar precision.

RAUCH, TREIMER and BONSE /3.22/ used a plane piece of material behind the second crystal slab to cause the phase shift. This plate included both partial beams I and II, see Fig.3.6. Turning the plate resulted in a continuous variation of the optical path differences ΔD between the two beams. The corresponding intensity-modulation of the outgoing beams 0 and H for an aluminium plate is shown in Fig.3.7.

In a previous work /3.23/ the authors reported measurements on several metal plates, which yielded values for the scattering length of the elements with uncertainties of ± 1 % $\Delta b/b$ which are restricted by the experimental inaccuracy in determining the wavelength ($\lambda = 2.04 \pm 0.02$ Å). Much higher accuracies, however, are attained most recently by the use of a more elaborate interferometer set-up at a high neutron flux facility /3.24/.

Summarizing, it should still be mentioned that a neutron optical inhomogeneous structure in the sample will destroy the phase relation between the partial beams. This fact reduces the applicability for the determination of scattering lengths but it makes possible the investigation of the inhomogeneities themselves /3.25/.

3.6 Mirror Reflection

For most elements the scattering length b is positive in sign so that the index of refraction becomes smaller than unit. Then a monochromatic neutron radiation incident on a surface is totally reflected if the glancing angle is smaller than the limiting angle

$$\gamma = (1-n^2)^{1/2} = \lambda(N\bar{b}/\pi)^{1/2} \tag{3.9}$$

with N being the atomic density (atoms per cm^3) of the bulk material. The limiting angle is determined by the average scattering length $\bar{b}$ because the neutron wave interacts with the whole medium rather than with individual atoms. Therefore, the limiting or critical angle could be measured for a mixture of substances (liquid, solid or gas) and the coherent scattering length obtained would be an accurate average of the scattering length of the constituents, regardless of their molecular form or crystalline state. Even inhomogeneities due to magnetic domains (about 10^{-3} cm in size), for instance, do not affect the critical angle for an unmagnetized mirror. Furthermore, the scattering length obtained from mirror reflection according to (3.9) is not affected by temperature vibrations of the atoms. Hence, there is no Debye-Waller correction to be made /3.26/ as in the diffraction and the fringe method.

The first reflection experiment with monochromatic neutrons was carried out by FERMI and MARSHALL /3.27/. HUGHES and BURGY /3.28/ used a neutron beam filtered through polycrystalline beryllium oxide or graphite.

These filters allow neutrons to pass through if their wavelength is longer than the wavelength of the Bragg cutoff. Because of the wavelength spread of the filtered beam, there is a considerable rounding of the reflection curve (reflected intensity versus angle) of a mirror around the limiting angle. Furthermore, it is nearly impossible to obtain an exact effective mean wavelength for an actual wavelength distribution. Therefore, the accuracy for scattering length obtained from direct determinations of the critical angle is restricted to the order of several percent $\Delta b/b$.

However, the reflection method is capable of greater accuracy when used to compare one scattering length with another. Using a Be-filtered reactor neutron beam and liquid mirrors of various hydrocarbons, DICKINSON et al. /3.29/ have measured in a very careful work the ratio of the scattering lengths of carbon to hydrogen with an accuracy within ± 0.2 %. A detailed discussion of how the accuracy of the measured scattering length depends on the spectral distribution of the incident filtered neutron beam is given in /3.30/.

A number of investigators utilized the mirror technique to measure the refractive index on solid mirrors. HEINDL et al. /3.31/ have shown that even mirrors consisting of evaporated separated isotopes on vanadium plates can be employed. BALLY et al. /3.32/ studied the reflection of a monochromatic neutron beam from metallic mirrors especially considering the depth of penetration of neutrons into the mirror substance. Examples for measured reflection curves on glass and copper and cadmium mirrors are shown in Fig.3.8. The accuracy of the determined scattering length was estimated to be within ± 2 % $\Delta b/b$.

To determine the index of refraction of gases, McREYNOLDS /3.33/ studied the reflection of a filtered reactor neutron beam from a liquid-gas interface and considered the decreasing reflectivity with increasing gas pressure. This procedure resulted in a value for the scattering length of the gas in terms of the mirror scattering length. Thus, the first data for noble gases (He, Ar) were determined.

Exact values for the scattering lengths of heavy noble gases are of high importance for the investigation of the fundamental neutron-electron interaction (see Section 4.4). For solving this problem CROUCH et al. /3.34/ carried out reflection measurements on the surfaces of the liquified gases and, for reference, on mixtures of ordinary and heavy water. The uncertainties of the results lie within ± 3 % $\Delta b/b$.

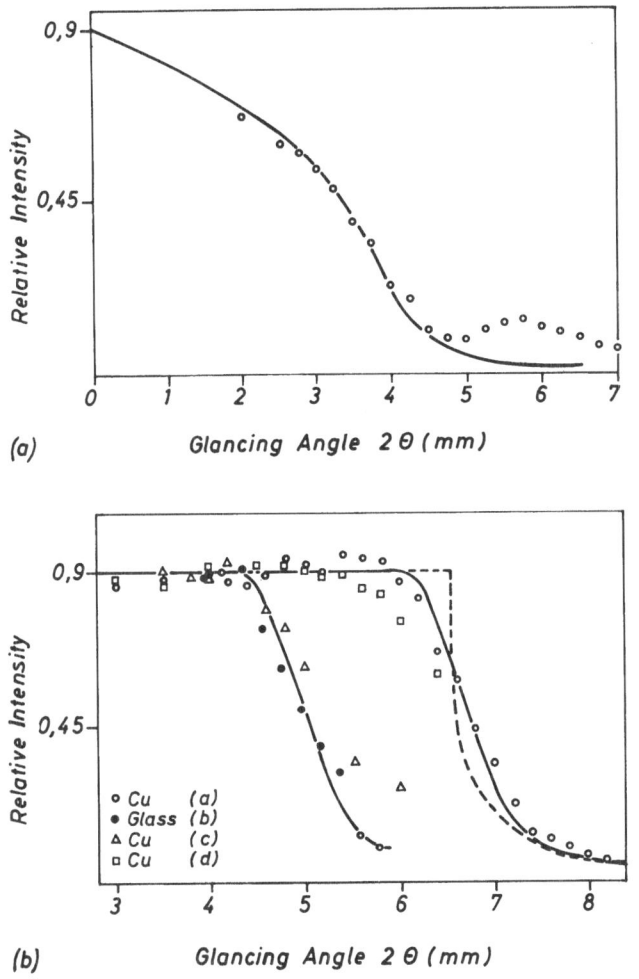

(a)

Relative Intensity

0,9

0,45

0   1   2   3   4   5   6   7

Glancing Angle 2Θ (mm)

(b)

Relative Intensity

0,9

0,45

○ Cu   (a)
● Glass (b)
△ Cu   (c)
□ Cu   (d)

3   4   5   6   7   8

Glancing Angle 2Θ (mm)

Fig.3.8 a and b. Reflection coefficient. (a) for cadmium, (b) for glass and various thick Cu layers on glass. Measured with monochromatized neutron beam. The dashed line in b gives the calculated reflection curve for a copper mirror and monochromatic neutrons /3.32/

Summarizing, one may state that the described mirror technique has a wide field of applications for solids, liquids and gases. The inherent accuracy is excellent because of the averaging nature of the reflection process. Also the angle and intensity measurements are simple in principle. A serious restriction of the accuracy is due to the wavelength spectrum of the neutron beam and to the uncertainty in the knowledge of the wavelength. Therefore, an improvement in the experiments must involve a method that would avoid any dependence on an accurate knowledge of the wavelength and its distribution of the incident neutrons. This was achieved by the method described in the following section.

3.7 Neutron Gravity Refractometer

3.7.1 Principle

In 1962 MAIER-LEIBNITZ /3.35/ established a unique principle for absolute and highly precise measurements of scattering lengths. The idea of Maier-Leipnitz is based on the fact that freely falling neutrons gain in the gravitational field an energy mgh, which is of the same order as the potential energy $<V>$ of neutrons in matter. h denotes the height of fall, g is the gravitational acceleration and m the neutron mass. Falling neutrons cannot penetrate an horizontal plane surface of matter, so long as the energy of the free fall mgh remains smaller than $<V>$. Then the neutrons are totally reflected. A critical height h_o for total reflection is reached when mgh equals $<V>$. In this case holds

$$mgh_o = <V> = 2\pi\hbar^2 m^{-1} N \cdot b. \tag{3.10}$$

If the height of fall exceeds h_o neutrons will penetrate into the mirror substance (N atoms/cm^2). The fundamental equation (3.10) for the reflection of neutrons in the gravity refractometer has some remarkable features. It exhibits the direct proportionality between the scattering length b and the critical height of fall h_o, which is the only quantity to be measured for determination of Nb and hence of b itself. All other quantities in the equation are very well known fundamental constants. Thus the basic equation allows the determination of b on an absolute scale with a high accuracy which is limited practically only by the experimental errors of the measurements of h_o and N. No neutron wavelength must be measured and no Debye-Waller factor must be taken into account.

When considering the order of magnitude we learn that the scattering length of the order 10^{-12} cm is measured on a scale of fall heights in the order 10^2 cm. This means an enlarging of one part in 10^{14}.

Furthermore, (3.10) gives a relation between the gravitational mass of the neutron in the term mgh and the inertial mass as quantity in the expression for the potential energy $<V>$. Therefore, the assumption of the equality of graviational and inertial mass for the free neutron must be made if scattering lengths are to be determined according to the basic equation (see section 4.3).

3.7.2 Arrangement

A diagram of the gravity refractometer built up /3.36, 37/ at the FRM reactor (4MW$_{th}$, open pool) is shown in Fig.3.9. Neutrons, emerging from the reactor in a nearly horizontal direction, fall under the action of gravity. They travel in an evacuated

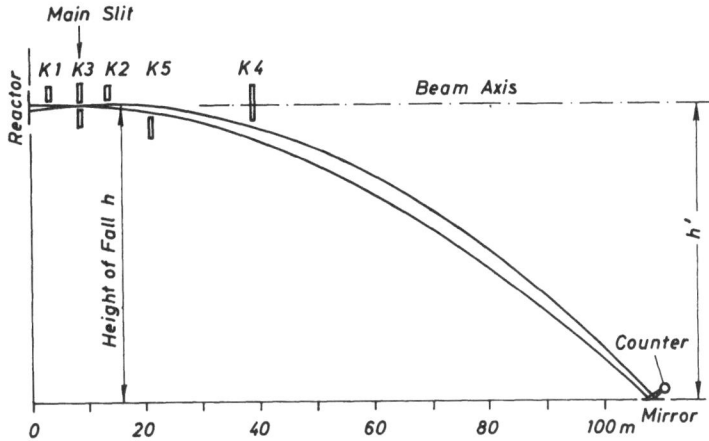

Fig.3.9. Principle of the neutron-gravity refractometer. K1...K5 are slits and stopper for the neutron beam /3.38/

tube of about 110 m in length in which they can reach heights of fall up to 170 cm.
The neutron path and the vertical momentum p_z of neutrons after a certain height of
fall h are dictated by gravity and centrifugal and Coriolis acceleration. This leads
to an energy gain in the vertical direction /3.38/ according to

$$p_z^2/2m = mg(h-\delta-c\cdot\sqrt{h}) = mgH \qquad (3.11)$$

where $\delta = l^2/2r$ is the difference in height between the horizontal line and the "surface" of the globe (radius r) at a distance l from the starting point. c denotes a
geometrical constant taking account of the Coriolis acceleration. The sum of the
correction terms amounts to -0.9% $\Delta H/H$ at h = 20 cm and -0.2 % $\Delta H/H$ at h = 150 cm.
The expression for the effective height of fall H itself is accurate within $\pm 10^{-8}$
$\Delta H/H$ /3.39/.

3.7.3 Neutron Reflection

The neutron paths from the source to the mirror, positioned at a distance h below
the main slit, have culmination points within a height region Δh which is limited to
about 10^{-3} h. This spread of fall-height causes a calculable rounding of the theoretical reflection curve which otherwise has a sharp cut-off at the critical height
h_0. Furthermore, the reflection curve is markedly rounded by absorption, incoherent

and inelastic scattering of neutrons by the mirror substance. If these factors are taken into account the reflectivity R for an exactly horizontal mirror is given by /3.37/

$$R(h_o,h,A) = \left\{\left[1-(1-h/h_o + iA/h_o)^{1/2}\right]/\left[1+(1-h/h_o + iA/h_o)^{1/2}\right]\right\}^2 \qquad (3.12)$$

The parameter A is equal to $N(\sigma_a+\sigma_i)(h\cdot1/4m):\sqrt{2/gh_o}$, where σ_a,σ_i are the cross sections for absorption and all kinds of incoherent scattering, respectively. In most cases, the influence of the fall-height spread on the reflectivity according to (3.12) may be neglected. Measured reflection curves on liquid mirrors of C_2Cl_4 and D_2O are shown in Fig.3.10 and 3.11. One can see a very good agreement of the experimental points with the calculated curves /3.37, 38/.

3.7.4 Accuracy

The inherent accuracy of the gravity mirror technique can be realized only with a liquid mirror whose surface is exactly in the horizontal plane. Liquid surfaces exhibit the best surfaces if they can be kept absolutely smooth. Great effort was made to fulfill this condition and to measure the fall heights and to determine the critical height with high precision /3.37/. NÖCKER /3.40/ has shown that even for mirrors of molten metals (Bi) an accuracy within $\pm2\cdot10^{-4}$ $\Delta b/b$ for the scattering length could be obtained. The measured reflectivities for liquid Bi, shown in Fig.3.12 agree very well with the calculated reflection curve corrected for the finite full height distribution.

A compilation of the uncertainties is given in Table 3.1. The most poorly known quantities are often purity and composition of the mirror substances. This is of some importance when mirrors consist of compounds of different elements. In a recent work /3.38/ it could be demonstrated that by the use of various compounds and by careful analysis of the compounds accuracies indeed between $\pm1\cdot10^{-4}$ $\Delta b/b$ and $\pm4\cdot10^{-4}$ $\Delta b/b$ for the scattering length of the compound elements can be achieved.

3.8 Christiansen Filter

In 1884 CHRISTIANSEN /3.41/ reported on experiments in which he studied the scattering of light by mixtures of powders or hairs with liquids. He has found that scattering disappears if the refractive index of the solids equals that of the liquid phase. When a well-collimated neutron beam is passed through such a Christiansen-filter containing a mixture of powder with liquid, the width of the beam is increased by refraction and diffraction by the individual particles. This small angle scattering was first discussed fully by WEISS /3.42/ who used it to determine the phase and scattering length of the powder by measuring the degree of broading

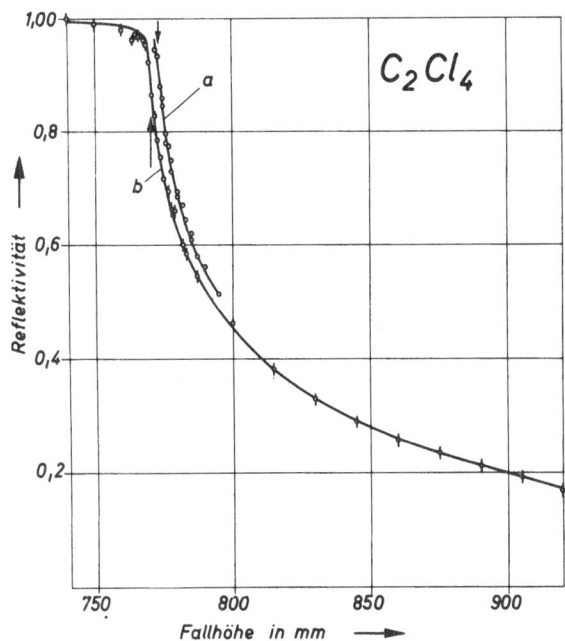

Fig.3.10. Reflection curves
of C_2Cl_4-mirror. The full
lines are calculated curves.
Curve a is measured at 20^0 C
under air, curve b at 26^0 C
under He-gas /3.36/

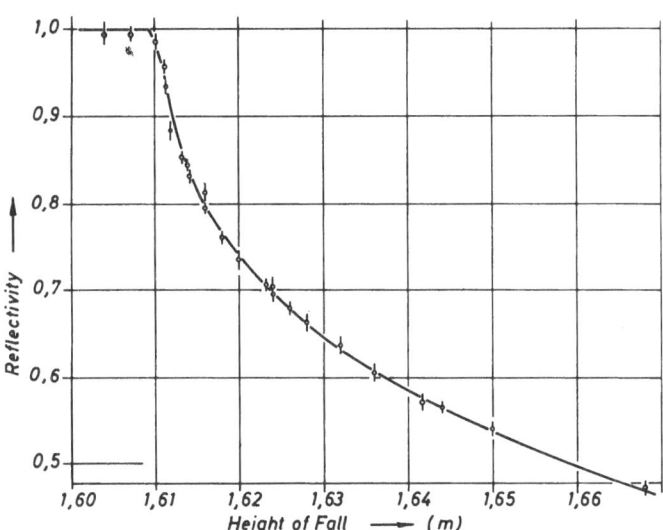

Fig.3.11. Reflection curve of heavy water (99.63 mol % D_2O). The full line is the
calculated curve for h = 1610 mm /3.53/

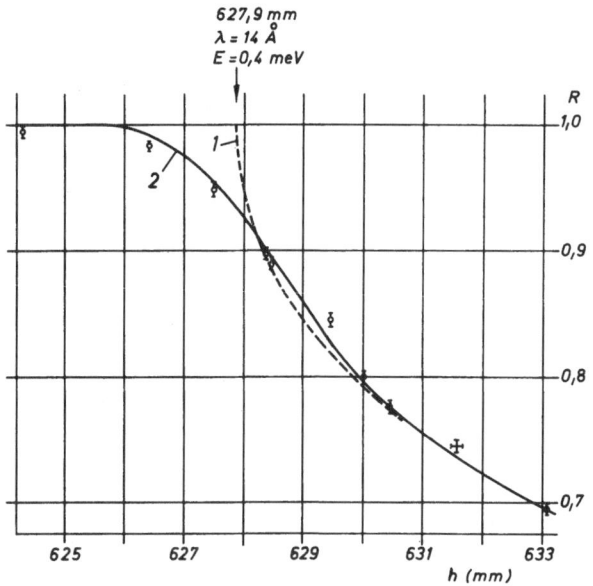

Fig.3.12. Reflection curve of liquid bismuth mirror (280° C). The dashed line 1 is calculated without fall height spread and absorption. The full line 2 is the calculated curve including all corrections

Table 3.1. Compilation of uncertainties in the determination of scattering lengths by means of the gravity refractometer

Acceleration by gravity g	$\pm 1 \cdot 10^{-5}$	$\Delta g/g$
Mass of neutron (squared)	$0.6 \cdot 10^{-5}$	$\Delta m^2/m^2$
Planck number (squared)	$1.2 \cdot 10^{-5}$	$\Delta \hbar^2/\hbar^2$
Avogadros number	$0.1 \cdot 10^{-5}$	$\Delta L/L$
Atomic weight	10^{-4} to 10^{-6}	$\Delta A/A$
Critical height H	$0.6 \cdot 10^{-4}$	$\Delta H/H$
Leveling	$0.4 \cdot 10^{-4}$ to $5 \cdot 10^{-4}$	$\Delta h/h$
Density	10^{-4}	$\Delta g/g$
Temperature	10^{-4}	$\Delta T/T$
Purity, sample composition	$(\sim 0.1 \cdot 10^{-4})$	$\Delta b/b$
Over all average	$(2..3 \cdot 10^{-4})$	$\Delta b/b$

of the beam. For neutrons the balance technique of Christiansen was first used by KOESTER and UNGERER /3.43/ to measure the scattering lengths of tungsten powder and heavy water in terms of the scattering length of reference liquids well known from reflection experiments in the gravity refractometer.

KOESTER and KNOPF /3.44/ made a detailed study of the small angle scattering process and the balance technique with cold neutrons in an arrangement sketched in Fig.3.13. The filters contained powders of elements. The variation of the scattering density $(Nb)_F$ of the liquid phase was achieved by suitable mixing of two known liquids, such as C_2Cl_4 and C_8H_{10} or H_2O and D_2O which cover a large range of the scattering density, ranging from $Nb = 0.55 \cdot 10^{16}$ cm^{-2} to $+6.38 \cdot 10^{10}$ cm^{-2}. Other organic liquids and mixtures can also be used.

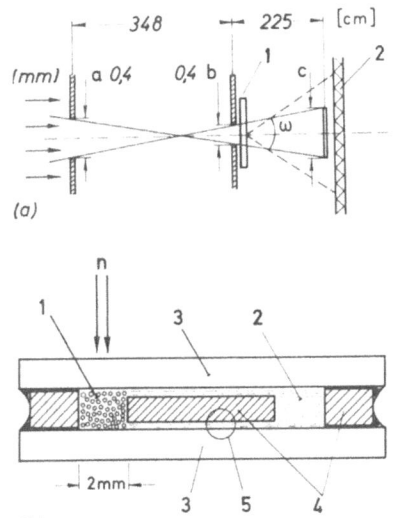

Fig.3.13 a and b. Experimental set-up for Christiansen filter measurements. (a) beam collimation: a,b slits; c stopper, 1 filter, 2 neutron counter. (b) cross cut of the Christiansen filter: 1 powder, 2 liquid, 3 glass, 4 distance holder, 5 liquid split /3.44/

In the arrangement shown in Fig.3.13, a beam of cold neutrons (effective wavelength 13 Å) emerging from a neutron guide /3.4/ at the FRM reactor enters a system of two narrow slits and hits the detector after having passed the filter. The unscattered part of the beam can be completely stopped by a Cd-shield positioned before the detector. Neutrons scattered by the filter into small angles are able to pass the stopper and to strike the detector. The ratio of the scattered neutrons to the total number measured without stopper represents a measure for the relative power of the small angle scattering. The variation of this figure with the scattering density of the liquid is shown in Fig.3.14 for filters of Al, Sb and Te. At the balance point in the minimum of the curves the wanted scattering density $(Nb)_p$ of the powder equals that of the liquid. Hence it holds

$$(Nb)_p = (Nb)_F. \qquad (3.13)$$

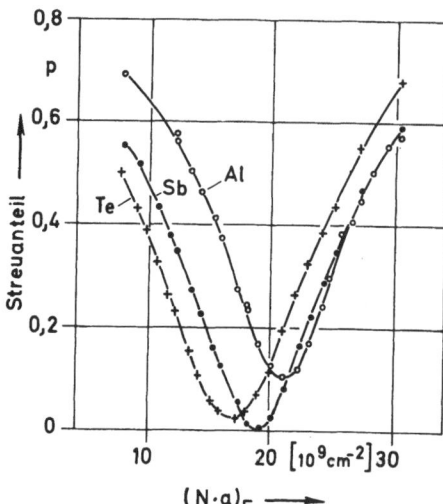

Fig.3.14. Small angle scattering by Al, Sb and Te powders in Christiansen filters /3.44/

The Nb-value at the balance point can be found with an uncertainty of about $\pm 3 \cdot 10^{-13}$ cm^{-2} which leads for b when also account is taken of the error in N to an accuracy within the range from ± 0.3 % to ± 1 % $\Delta b/b$. In principle, there is no systematic error. But a water content of the powder, dissolved gases, impurities and deviations from stochiometry of compounds cause uncertainties which must be carefully considered.

The Christiansen filter technique is, of course, not limited to powder-liquid mixtures. Combinations of thin wires with liquids /3.45/ and of powders with gases under various pressures /3.46, 47/ were successfully applied.

Concluding, one may state that the Christiansen filter technique is a very useful and reliable method for the determination of scattering lengths with medium accuracy within 0.1 % to 1 % $\Delta b/b$. It is applicable to elements, single isotopes /3.48/ and chemical compounds /3.49/ in the form of powders, liquids or gases.

3.9 Transmission

3.9.1 Introduction

Neutron transmission experiments at eV energies provide data for the total cross
section from which the free scattering cross section or the potential scattering
cross section /2.5/ can be accurately derived if the binding effects become negligible and the contributions of neutron capture or resonance scattering are small
or can be calculated. Accurate knowledge of the total scattering cross section is
always then of great interest when exact experimental data of coherent scattering
lengths or incoherent cross sections are available too. In these cases, the fundamental spin state scattering lengths (see Section 2.2.2) for $I \neq 0$ mono-isotopes,
or another unknown quantity, the scattering length or the incoherent cross section
can be determined. Even the small contribution of the neutron-electron interaction
to the atomic cross section can be estimated by very precise measurements of cross
sections and scattering lengths.

   In order to achieve a good accuracy for the derived quantities the cross sections
must be determined with an accuracy comparable to the high precision (within some
10^{-4} $\Delta b/b$) of the scattering lengths. Although the transmission method is simple
in principle and though the experiments are not very complicated, the compilation
of nuclear data /3.50/ shows a lack of exact data for this quantity.

   In the last decade great effort was made to measure accurately the free scattering cross sections of the (n,p) /3.51, 52/ and (n,d) /3.53/ interactions which
are of great interest in nuclear physics, and also the free cross section for a few
standard nuclei which show no or totally incoherent scattering, such as carbon
/3.51, 54/ or vanadium /3.55/, respectively.

3.9.2 Experimental

Neutron sources for experiments with collimated beams of neutrons with eV-energies
provide a moderated spectrum of a fast neutron flux produced by fission reactors
or accelerators. For the selection of the wanted energies, various techniques were
developed, such as producing of monochromatic beams by single /3.27/ or double /3.56/
Bragg reflection, by filtering the beam /3.57/ or by selecting of neutrons according
to their velocity /3.58 - 60/ or by resonance absorption /3.57, 62/ of neutrons of
the corresponding resonance energy.

   The requirements of high precision measurements, high and stable neutron intensity, and sharp energy selection, could be fulfilled by using filtered reactor neutron beams and applying the resonance scattering /3.54/ and resonance detector /3.63/
techniques. DILG and VONACH /3.64/ studied the double resonance scattering technique.

A nearly mono-energetic neutron beam is produced by means of a first resonance scatterer (W: 19 eV; Co: 130 eV and Na: 2.8 keV) near the reactor core. After a flight path of about 7 m this beam is scattered into a neutron detector by means of the second resonance scatterer identical with the first one. A scheme of the experimental arrangement is shown in Fig.3.15. Because of the double scattering, only a very small fraction of nonresonant neutrons may strike the counter. Even this small contamination can be eliminated by removing the resonant neutrons by a resonance filter and measuring the rate of the remaining nonresonant scattering.

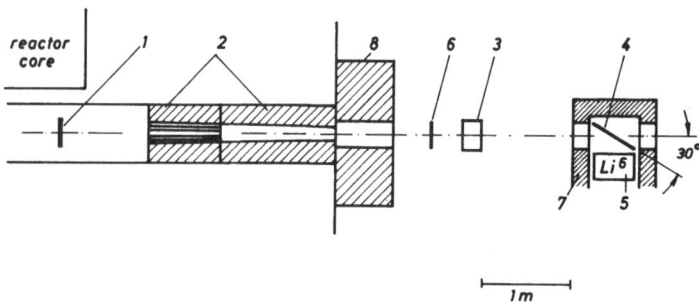

Fig.3.15. Schematic drawing of the double resonance-scattering arrangement. 1 first scatterer: Co-foil; 2 beam collimator; 3 sample; 4 second scatterer: Co-foil; 5 Li^6 J(En) detector; 6 Co-filter for difference measurements; 7, 8 shielding /3.54/

KOESTER and SCHACHT /3.63/ made a detailed study of the resonance detector technique. They proposed a continuously operating resonance detector consisting of a thin rotating foil of a resonance absorber (Rh: resonance energy E_r = 1.2 eV or Ag: E_r = 5.2 eV), which is activated at its periphery by the neutron beam, as shown in Fig.3.16. Diametrically to the activation spot the β-activity of the foil periphery is continuously measured with β-counters. Thus a long measuring time, not limited by the half-life of the activity, and hence a higher statistical accuracy can be achieved. In a precision transmission experiment WASCHKOWSKI and KOESTER /3.65/ used a combination of two identical rotating detectors, as sketched in Fig.3.17. The first detector foil is activated by the original neutron beam giving a measure for the sum of resonant and nonresonant activation, whereas the second foil is activated only by the nonresonant neutrons able to pass through the first foil. Thus, the difference of the two simultaneously measured activities is due only to the resonant neutrons.

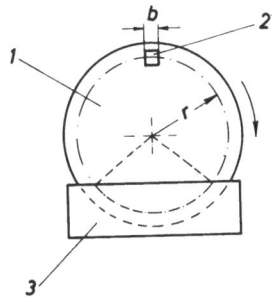

Fig.3.16. Schematic drawing of the rotating resonance-
absorber device. *1* rotating foil; *2* cross section
of neutron beam; *3* β-counter. r = 10 cm, b ≈ 1 cm
/3.63/

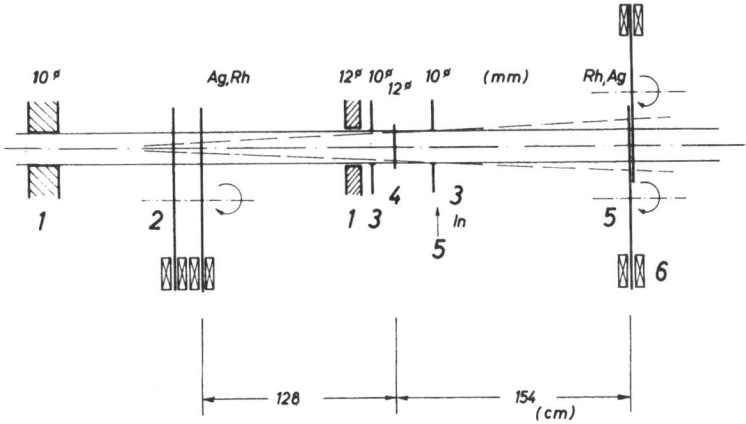

Fig.3.17. Arrangement for transmission measurements with rotating resonance detectors. *1* shield; *2* detectors as monitor; *3* slit; *4* sample; *5* detectors; *6* β-counters

Moreover, the ratio of the two activities is a measure for the energy distribution of the neutrons so that a variation of the distribution caused by the transmission sample can be detected. On the other hand, a variation might indicate an
energy-dependent cross section.

In experiments with solid and liquid samples of lead and bismuth the authors
could show that at 1.2 eV and 5.2 eV neutron energy the accuracy of the transmission method was limited to the order of ±0.2 % $\Delta\sigma/\sigma$ for solid samples. The limitation was caused by inhomogeneities in crystal structure and density of the
samples. With liquid samples of the same elements they have, however, achieved an
accuracy within ±5·10^{-4} $\Delta\sigma/\sigma$ in accordance with the statistical error. This is now
of the same order as the accuracy of the best determined scattering lengths.

4. Experimental Results

4.1 Scattering Lengths

So far, we have seen the considerable efforts made to determine scattering lengths
with high accuracy. The techniques used were different in both physical principle
and in experimental performance so that the question arises how good the accordance
of the results is.

Data obtained by the different techniques are compiled in Table 4.1. The first
group (column 2) includes examples for original data from diffraction work and from
the refined reflection experiments. The results are given at last in terms of the
carbon scattering length which was deduced from the measured free-scattering cross
section. In the other groups, results of experiments are given in which the advanced
methods of the dynamical diffraction and interferometer experiments (column 3), the
gravity-mirror and the Christiansen filter techniques (column 4) were applied. These
values and the results of the transmission experiments (column 5) have been determined on an absolute scale. Considering the given figures one observes good agreement within the limit of error between the data of the transmission experiments and
of the advanced methods. Also the accordance with the data from reflection experiments is excellent, whereas the comparison with the diffraction results suffers
from the large errors in the latter results. In many other cases, not shown here,
a close accordance, however, between more accurate diffraction data and results
from Christiansen filter measurements can be observed. This accordance demonstrates
the reliability of the errors in the order of 10^{-3} to 10^{-4} $\Delta b/b$ as given for the
advanced methods. This order of accuracy makes it possible that differences of the
order of 1% $\Delta b/b$ can be determined within an accuracy of 10% to 1%. We will see in
the following sections how great the importance of this capability is.

4.2 Incoherent Cross Sections

Exact measurements of total cross sections and exact values of the scattering length
a_N provide the incoherent cross section for free nuclei according to

$$\sigma_{inc}^{free}(0) = \sigma_{tot}(E) - \sigma_{abs}(E) - \sigma_K(E) - 4\pi a_N^2(0) = \sigma_o(free) - 4\pi a_N^2(0) \qquad (4.1)$$

where σ_K denotes a correction term which is mainly due to solid-state effects and
contributions of shell electrons and nuclear resonances. For the use in applications
of slow neutrons we give in Tabe 4.2 a list of scattering lengths, of derived
values (with $b_{ne} = -1.38 \cdot 10^{-3}$ fm) for $4\pi a_N^2$, of free cross sections at zero energy

Table 4.1. Scattering lengths obtained by different techniques (in fm)

1	2	3	4	5

	D Diffraction
	R Reflection
		DD: Dynamical
		    diffraction
		J: Interfero-
		    meter
			G Gravity-re-
			   fractometer
			Ch Christiansen
			   filter
				Transmission

H	D -4.0(1) R -3.74(2)	--	G -3.7409(11)	-3.733(4)
D	D 6.4(3) R 6.21(4)	--	G 6.674(6) Ch 6.70(5)	--
C	(reference standard) ^{a}	--	G 6.6484(13)	6.652(6)
O	R 5.80(5) D 5.81(20)	Pr^c 5.830(2)	G 5.803(5)	5.804(7)
Al	D 3.5(2)	J 3.447(5)	Ch 3.449(9)	3.455(5)
Si	--	DD 4.1491(10) Pr 4.1478(16)	Ch 4.159(6)	4.152(9)
Cl	D 9.9(2)	--	G 9.5792(8)	--
Ge	D 8.4(2)	DD 8.1858(36) Pr 8.1929(17)	--	--
Hg	R 12.58(13)^b	--	G 12.66(2)	--
Bi	D 8.9(3)	J 8.580(8)	G 8.5256(14)	8.5267(16)

a Reference value: b(c) = 6.65(2) fm.
b Based on b(C) = 6.6484(13) fm and b(H) = -3.7409(11) fm.
c Prism refraction, values taken from /3.66/.

Table 4.2. List of scattering lengths (in fm), free and incoherent cross sections for elements (in barns)

Z	El.	b (recommended) 1.	b (recommended) 2.	$4\pi a_N^2$	σ_0(free)	σ_{inc}(free)
1	H	-3.7409(11)	-3.733(4)	0.4390(3)	20.491(14)	20.052(15)
	D	6.674(6)	6.73(18)	2.486(5)	3.390(12)	0.904(13)
	T	5.0(3)				
2	He	3.0(2)		0.72(10)	0.773(9)	0.05(10)
3	Li	-2.03(5)	-1.94(5)	0.39(2)		
4	Be	7.8(4)		6.2(6)	6.151(5)	0.004(1)[a]
5	B	5.35(6)	5.40(4)	3.02(7)	3.62(11)	0.6(2')
6	C	6.6484(13)		4.7391(19)	4.7461(45)	0.007(8)
7	N	9.36(2)	9.40(15)	9.60(5)	10.03(8)	0.4(1)
8	O	5.803(5)	5.830(2)	3.759(8)	3.761(7)	0.002(11)
9	F	5.66(2)	5.603(11)	3.646(26)	3.575(14)	0.0004[b]
10	Ne	4.6		2.4	2.42(3)	0.0(1)
11	Na	3.63(2)		1.524(17)	3.12(1)	1.60(3) 1.36(13)[b]
12	Mg	5.38(2)		3.375(25)	3.416(4)	0.041(26) 0.071(6)[a]
13	Al	3.449(9)	3.447(5)	1.404(8)	1.4082(11)	0.0071[b] 0.0098[a]
14	Si	4.1491(10)		2.0350(10)	2.0420(20)	0.007(3) 0.014(2)[a]
15	P	5.13(1)		3.128(12)	3.134(10)	0.06(16)
16	S	2.847(1)		0.9725(8)	0.985(4)	0.011(4)
17	Cl	9.5792(8)		10.956(2)	16.6(3)	5.6(3)
18	Ar	1.8(2)	2.0(2)	0.38(7)	0.638(6) 0.647(3)	0.26(11)
19	K	3.71(2)		1.667(18)	2.03(10)	0.36(10)
20	Ca	4.90(3)	4.88(7)	2.90(3)	2.90(3)	0.00(6)
21	Sc	12.15(13)	12.8(2)	17.8(3)	22.1(4)	4.26(22)

Table 4.2 (continued)

Z	El.	b (recommended) 1.	2.	$4\pi a_N^2$	σ_0(free)	σ_{inc}(free)
22	Ti	-3.37(2)	-3.4(2)	1.34(2)	4.09(3)	2.75(4) 2.60(21)[a]
23	V	-0.408(2)	-0.42(1)	0.0171(2)	4.80(5)	4.78(5)
24	Cr	3.532(10)	3.52	1.54(2)	3.37(2)	1.83(3)
25	Mn	-3.73(2)	-3.7(1)	1.65(2)	2.2(2)	0.6(2)
26	Fe	9.54(6)	9.16(13)	11.12(14)	11.33(4)	0.21(15) 0.38(3)[a]
27	Co	2.78(4)	2.50(5)	0.964(28)	6.00(6)	5.04(8) 4.7(2)[b]
28	Ni	10.3(1)		12.98(26)	17.8	4.8(6)
29	Cu	7.689(6)	7.63(4)	7.274(12)	7.74(4)	0.47(5) 0.50(4)[a]
30	Zn	5.7(1)		4.03(14)	4.05(6)	0.02(16) 0.075(7)[a]
31	Ga	7.2(1)		6.3(1)	6.5(2)	0.2(3)
32	Ge	8.1858(36)	8.1929(17)	8.281(8)	8.37(6)	0.09(10)
33	As	6.73(2)	6.4(1)	5.93(4)	6.7(7)	0.8(8)
34	Se	7.95(4)	8.10(5)	7.84(8)	8.1	0.26
35	Br	6.77(2)	6.79(2)	5.70(4)	5.8(4)	0.1(4)
36	Kr	7.91(15)	7.68(19)	7.77(30)	7.61(4)	
37	Rb	7.08(2)	7.04(8)	6.24(4)	6.2(3)	0.0(4)
38	Sr		6.88(13)			
39	Y	7.75(2)	7.65(7)	7.48(4)	7.60(6)	0.12(8) 0.15(1)[a]
40	Zr	7.0(1)		6.12(18)	6.23(1)	0.11(18)
41	Nb	7.11(4)	7.08(2)	6.32(7)	6.37(4)	0.0062(6)[a] 0.0024(3)[b]
42	Mo	6.95(7)	6.4(2)	6.06(12)	5.97(4)	
43	Tc	6.8(3)		5.8(6)		
44	Ru	7.21(7)	7.3(1)	6.5(1)	6.4(1)	0.0(1)
45	Rh	5.88(4)	5.91(4)			

Table 4.2 (continued)

Z	El.	b (recommended) 1.	2.	$4\pi a_N^2$	σ_0(free)	σ_{inc}(free)
46	Pd	6.0	6.4(3)			0.091(9)[a]
47	Ag	6.02(2)	6.1(2)	4.57(3)		0.52(5) 0.48(4)[a]
48	Cd	3.8 + i1.2				
49	In	4.08(4)	3.9(1)			
50	Sn	6.220(2)	6.1(1)	4.887(3)	4.88(2)	0.00(6) 0.022(5)[a]
51	Sb	5.641(12)	5.4(1)	4.03(2)	4.2(1)	0.17(12)
52	Te	5.43(4)	5.6(2)	3.74(6)	4.4(4)	0.6(4)
53	J	5.28(2)	5.25(4)	3.54(3)	3.5	~0
54	Xe	4.88(3)	5.22(11)			
55	Cs	5.42(2)		3.74(2)	~8.39	~4.6
56	Ba	5.28(5)	5.22(13)	3.55(7)	3.42(4)	
57	La	8.27(5)	8.32(14)	8.63(10)	9.5	~1 1.84(17)[a]
58	Ce	4.83(4)	4.82(6)	2.99(5)		
59	Pr	4.45(5)	4.4(4)	2.54(6)		
60	Nd	7.80(7)	7.2(2)	7.70(14)		11(2)[a]
65	Tb	7.38(3)	7.56(20)	6.92(15)		
66	Dy	17.1(3)	16.9(4)	36.7(7)	34(7)	
67	Ho	8.5(2)		9.2(5)	9.5(2)	
68	Er	8.03(3)	7.9(2)	8.19(7)	11.0(8)	
69	Tm	7.05(5)	6.9(2)	6.3(1)	12(2)	
70	Yb	12.62(12)	12.6(6)	20(2)	25.0(8)	
71	Lu	7.3(2)		6.8(4)	8(2)	
72	Hf	7.77(14)	8.8			
73	Ta	6.91(7)	7.0(3)	6.11(13)	6.66	0.020(4)[a] 0.011(3)[b]
74	W	4.77(5)	4.66(4)	2.95(6)	5.0	~2 1.84(12)[a]

Table 4.2 (continued)

Z	El.	b (recommended) 1.	2.	$4\pi a_N^2$	σ_0(free)	σ_{inc}(free)
75	Re	9.2				
76	Os	10.8				
77	Jr	10.6(2)	10.0(2)		14	
78	Pt	9.5(3)		11.5(8)	11(1)	0.59(4)[a]
79	Au	7.63(6)	7.7(1)	7.45(12)		0.36(4)[a] 0.38(7)[b]
80	Hg	12.66(2)	12.67(13)	20.29(7)	~30	~10
81	Tl	8.89(2)		10.08(5)	9.7(4)	
82	Pb	9.4003(14)		11.263(4)	11.262(5)	0.001(7) 0.0013(5)[a]
83	Bi	8.5256(14)	8.580(8)	9.2907(33)	9.3001(31)	0.009(5) 0.0071(6)[a] ≤0.02 mb
90	Th	10.08(4)	10.52(6)	12.97(10)	12.67(8)	~0
92	U	8.61(4)	8.51(22)	9.422(88)		
93	Np	10.6				

a σ_{inc} directly measured /4.49/

b   Spin-dependent σ_{inc} determined by means of pseudomagnetic measurements /4.50/

and of deduced as well as directly measured incoherent cross sections. Two recommended values for the atomic scattering length are listed. The first one is believed
to be the best value at present, whereas the second one is an alternative value obtained by another technique. If the first value is unpublished the second one represents a published result. The value $4\pi a_N^2$ is always derived from the first
recommendation. It should be noted that the present incoherent cross sections differ
considerably in some cases from values previously reported on /4.1, 2/.

4.3 Neutron and Gravity

Gravitational acceleration and gravity are phenomena which have been studied with
meter-sized objects and by observation of the planetary system since the beginning

of modern science. The fundamental principle of equivalence of inertial and gravitational mass could be verified within stringent limits of 2 parts in 10^{12}. On the
other hand, there are only few and rather crude gravitational experiments with particles of atomic /4.3/ or nucleon-size /4.4 - 6/ and with electrons /4.7/ and photons
/4.8/. The graviational acceleration of free neutrons was investigated for the first
time by McREYNOLDS /4.4/ who observed the fall of a thermal and a filtered (BeO)
neutron beam with a lower mean velocity. He found g_f = 935 $\pm$ 70 cm/s^2 in agreement, within the experimental error, with the usual value 980 cm/s^2. DABBS et al.
/4.5/ determined the gravitational acceleration by measuring the difference in fall
of neutrons of different velocity but defined by the same collimating system. This
system allowed well-collimated beams of both high- and low-velocity neutrons to
traverse a very long (180 m) evacuated flight path. Boral filters permitted the
selection of the first beam, which did not fall appreciably. The transmission edges
associated with lattice spacings (d_{hkl}) in a polycrystalline beryllium filter were
used to define the neutron velocities v_{hkl} = $\pi\hbar/(md_{hkl})$ in the slow beam. The differences in fall between the fast beam and the two slow beams <100> and <002> gave
values for the gravitational acceleration g_f = 975 $\pm$ 3 cm/s^2 and g_f = 973 $\pm$ 7 cm/s^2
which are slightly lower by $(g_f-g)/g$ = -(0.4 $\pm$ 0.3)%, than the local value g =
979.74 cm/s^2.

Another possibility to determine g_f is given by the fact that the value of a
scattering length b_f determined by means of the gravity refractometer depends on g_f.
By comparing the measured b_f with a value b obtained from measurements independent
of g one gets the relation g_f = g b_f/b. Precision measurements of the scattering
lengths of carbon using the gravity-mirror and the transmission method have given
the values listed in Table 4.1 which led to g_f = (0.9996 $\pm$ 0.0007)g = 980.35 cm/s^2
/3.37/. Most recent experiments on Pb and Bi /2.12/ and taking account of the neu-
tron-electron-interaction resulted in g_f/g = 1.00016 $\pm$ 0.00025. This means, as
shown by KOESTER /4.9/ a verification of the equivalence of inertial and gravitational mass for the neutron with an uncertainty of 3 parts in 10^4.

By the experiments of DABBS et al. /4.5/ and KOESTER /4.6/ it could be also shown
that the gravitational acceleration of the free neutron does not depend on the orientation of the neutron-spin to the vertical direction. No splitting of g_f greater
than $\Delta g_f/g_f$ = $\pm$0.5 % was found /4.6/. McREYNOLDS /4.10/ reported on resonance measurements on a neutron beam polarized up or down in the earth's field, which have
shown the gravity constants of both orientations to be equal within experimental
limits of only

$$\Delta g/g = \pm 5 \cdot 10^{-13}.$$

Using a perfect crystal interferometer as described in Section 3.5, COLLELA et
al. /4.11/ observed a gravitationally induced interference between two coherent

neutron beams having traversed the same distance but at different heights. By this experiment, the extremely high sensitivity of interference measurements with neutrons could be impressively demonstrated.

The results of the experiments with neutrons are rather crude compared with the high accuracy of 2 parts to 10^{12} of measurements on laboratory objects and massive bodies /4.12 - 15/ but they show that there is no experimental indication for a violation of the principle of equivalence for a free nucleon.

4.4 Neutron-Electron Interaction

4.4.1 Introduction

Neutron and electron have magnetic moments, and hence there exists the familiar magnetic dipole-dipole interaction between them and also an interaction between the magnetic moment of the neutron and the magnetic field associated with the convection current of an electron in motion. These interactions have been investigated in great detail by scattering slow neutron from magnetic ions and magnetic structures. They are not of interest here. A further electromagnetic interaction between the two particles is expected if regions with non-zero charge density exist inside the neutron. Any charged particle penetrating this charge distribution is then exposed to the action of electromagnetic forces. The interaction caused by the possible intrinsic property of the neutron is spin- and velocity-independent. By considering the interaction of two Dirac particles FOLDY /2.10/ has shown that the anomalous magnetic moment of the neutron implies a further contribution to the spin-independent (n,e) interaction, already presented as b_F in Section 2.2.3. Consequently, determination of the (n,e) scattering length $b_{ne} = b_F + b_e$ leads to a value for the intrinsic scattering length b_e. Accurate experimental data for b_{ne} are of fundamental interest in physics because they involve the internal structure of an elementary particle.

4.4.2 Experiments and Methods

Nearly all experimental studies up to now are based on measurements of the angle (θ)- or wavelength-dependence of the scattering amplitude for diamagnetic atoms, given by

$$b(\lambda,\theta) = b_N + Zf(\lambda^{-1}\sin\theta/2)b_{ne}.$$

FERMI and MARSHALL /4.16/ for the first time, and later KROHN and RINGO /4.17/ have observed the asymmetry of scattering of thermal neutrons from noble gases at different angles (45^0 and 130^0) due to the variation of f with angle. The asymmetry of (n,e) scattering shows up against a strong background due to the isotropic nuclear scattering, which requires calculable corrections that are about five times as large as the (n,e) effect.

A second method was employed by HAVENS et al. /4.18/ in which the dependence of the total scattering cross section $4\pi b^2(\lambda)$ in the region $\lambda \sim 1$ Å is observed. The nuclear scattering length should remain the same, whereas the formfactor $f(\lambda) = F(\lambda)$ from (2.9) is the main cause of the λ-dependence of the scattering cross section. The (n,e) effect in experiments on liquid Bi in the region from $\lambda = 0.3$ Å to $\lambda = 1.3$ Å by MELKONIAN et al. /4.19/ amounted to a variation of 1% of the total scattering cross section. Here the corrections as discussed in the preceding sections came to 40% of the (n,e)-contribution.

An ingenious idea was realized by HUGHES et al. /4.20/. The authors measured the critical angle for neutron total reflection on a parting plane between liquid oxygen and solid·bismuth. The nuclear scattering length density (Nb) of bismuth is only 2 % smaller than that of oxygen, whereas the (n,e) scattering contribution by bismuth is 9-times larger than that of oxigen. The measured value of the critical angle was almost equally determined by the (n,e) interaction and the noncompensated nuclear scattering. With the aid of separately measured data (at 10 eV energy) for the nuclear scattering the (n,e) scattering length could then be derived from the measured critical angle.

A relatively high (n,e) contribution of about 20 % was also obtained by ALEXANDROV /4.21/ by means of diffraction measurements on a single crystal of a tungsten isotope mixture. One of the isotopes, the W-186, has a negative scattering length of -0.47(1) fm while the other isotopes have positive scattering lengths of such magnitude, that an enrichment of W-186 to 84 % leads to a vanishing mean nuclear scattering length. The Bragg-reflected neutron intensity is then only defined by the interaction of the neutrons with the shell electrons. ALEXANDROV et al. /4.22/ carried out experiments on single crystals from two isotope mixtures enriched by W-186 and having nuclear scattering length of opposite sign. The results were not in agreement with each other, but it was shown that an interpretation of the results is possible by introducing a hypothesis about an additional scattering contribution into the Bragg peaks. So long as the existence and nature of the additional scattering remain unexplained, a discussion of the $b_{n,e}$ obtained should be omitted.

Most recently KOESTER et al. /2.12/ reported on precision measurements of the atomic scattering length b of lead and bismuth by means of the neutron gravity refractometer. Additional transmission measurements with eV-energy neutrons on molten lead and bismuth probes /3.65/ yielded exact values for the free cross sections from which the nuclear scattering lengths (b_N) could be derived (see Section 3.9). After taking account of all corrections which came to 4 % of the total (n,e) effect the neutron-electron contribution is given by

$$Z\, b_{n,e} = \left[b - b_N(\lambda) \right] / \left[1 - F(Z,\lambda) \right]$$

where $F(\lambda)$ is the atomic formfactor (2.9) for neutrons of wavelength λ with which the nuclear scattering length $b_N(\lambda)$ were measured. In the eV-region ($\lambda \sim 0.2$ Å) the formfactor for Pb and Bi amounts to $F(0.2$ Å$) \simeq 0.1$, so that a relatively high (n,e) effect is measured. In addition, a small uncertainty in the knowledge of $F(\lambda)$ introduces only a very small error. Furthermore, this method was applied to several light and medium elements in order to confirm that the technique is free of systematic uncertainties. The described experiments all exhibit an important example for the application of the advanced methods for the exact determination of scattering lengths and cross sections.

4.4.3 Results

The neutron scattering length $b_{n,e}$ is related to the slope $dG_{En}(0)/dQ^2$ of the electric formfactor of the neutron, which is measured in high energy scattering (n,d and n,p) experiments at various values Q^2 of the four-momentum transfer. At $Q^2 = 0$ we have

$$dG_{En}(0)/dQ^2 = dF_{1n}(0)/dQ^2 - F_{2n}(0)/4m^2 = -14.41(fm) \cdot b_{n,e}$$

F_{1n} and F_{2n} are the Dirac and Pauli form factors, respectively. The derivative of $F_{1n}(0)$ is a measure for the intrinsic charge distribution of the neutron whereas $F_{2n}(0)/4m^2 = -0.02115$ fm^2 corresponds to the Foldy term b_F described above.

The most exactly measured data for $b_{n,e}$ are compiled in Table 4.3, in which also information about the magnitude of the experimental (n,e) effect (column 2) and the applied corrections (column 3) are given. It is remarkable that the most exact values for $b_{n,e}$ were obtained by experiments with the smaller (n,e) contribution of the order of 1 % $\Delta b/b$. Also the relation between the necessary corrections (column 3) and the (n,e) effect was unfavourable for the earlier experiments. In spite of these facts, the results are compatible and agree with the mean of all values (-1.41 ± 0.06 fm) as well as with the most exact value ($-1.38 \pm 0.02 \cdot 10^{-3}$ fm) within a factor of two or three of the individual standard deviations (column 6). In particular, the mean with regard to the subsequently corrected data (column 5) is still consistent with the Foldy term, but the newest result of KOESTER et al. /2.23/ shows clearly a deviation of nearly 5 standard deviations from the Foldy value $-1.468 \cdot 10^{-3}$ fm. From the difference $b_e = (b_{n,e} - b_F) = +0.09 \pm 0.02 \cdot 10^{-3}$ fm a value for the derivative of the Dirac form factor

$$d F_{1n}(0)/dQ^2 = (-0.13 \pm 0.03) \ 10^{-2} \ fm^2$$

is obtained. This quantity indicates the existence of a small intrinsic charge contribution to the electrical structure of the neutron. As far as theoretical predictions are concerned SU(6), which works so well for the ratio of the magnetic

Table 4.3. Results of (n,e) experiments

Authors Year	Magnitude (relative)		$-b_{n,e}(10^{-3})$fm			Deviation (corrected)
	effect (n,e)/total	correction corr/total	reported	corrected	rel. error	
KROHN, RINGO 1973	0.5% $\Delta\sigma/\sigma$	3% $\Delta\sigma/\sigma$	1.30(3)	1.33(3)[a]	2.3%	+ 3.6%
MELKONIAN et al. 1959	1.3% $\Delta\sigma/\sigma$	0.6% $\Delta\sigma/\sigma$	1.56(5)	1.49(5)	3.3%	− 8%
HUGHES et al. 1953	~50% $\Delta\theta/\theta$	−	1.39(13)	−	10%	− 0.7%
average 1973			1.42(8)	1.40(5)	3.6%	− 1.4%
ALEXANDROV et al. 1974	~19% $\Delta b/b$		$\begin{bmatrix}1.06\\2.2\end{bmatrix}$	$[1.55]$[b]	−	− 12%
KOESTER et al.	1.2% $\Delta b/b$	0.05% $\Delta b/b$	1.38(2)	1.38(2)	1.5%	0
average 1976			1.41(6)	1.40(4)	2.4%	− 1.4%

a Subsequently corrected for Schwinger scattering and a resonance contribution.

b Corrected by authors corresponding to the hypothesis about additional magnetic scattering.

proton and neutron formfactors, they give the poor result $G_{En}(Q^2) = 0$ for all Q^2 /4.47/ and, hence $dG_{En}(0)/dQ^2 = 0$ in contrast to the experimental result $dG_{En}(0)/dQ^2 = -0.0199(3)$ fm^2. An improvement has been recently proposed by YAOUANC et al. /4.48/ who considered a breaking of SU(6).

4.5 Neutron-Proton Interaction

4.5.1 Introduction

The two-nucleon interaction is perhaps the most central problem in modern physics. MORAVCSIK /4.23/ has recently given an impressive and complete review of this im-mensely wide field. Among the numerous (n,n), (n,p) and (p,p) reactions in the energy range from zero to some GeV, the (n,p) interaction at lowest energies forms a sepa-rate subject for two main reasons: The experimental information in this region can be of very high accuracy, and furthermore, the analytical tools are different from those at higher energies because only very few angular momentum states are of importance.

Accurate knowledge of the (n,p)-scattering parameters at low energy is important for understanding and testing the validity of charge independence and the spin de-pendence of the nucleon-nucleon interaction. Using the effective range theory /2.3/, the (n,p) scattering cross section below 20 MeV can be described in the shape-in-dependent approximation in terms of only four parameters (4.24). These are the scat-tering lengths (a) and effective ranges (r_0) for singlet and triplet state, respec-tively. The scattering length $a_p = (3/4)a_t + (1/4)a_s$ is combined with the scattering cross section at "zero" energy: $\sigma_0 = 3\ a_t^2 + a_s^2$. The triplet effective range r_{ot} is determined by a_t and the deuteron binding energy, while the singlet effective range is derived from a_s and from the energy dependence of the total scattering cross section in the MeV-range (see e.g. NOYES /4.25/, HOUK /3.51/, LOMON and WILSON /4.26/ and DILG /3.52/.

In particular, the singlet effective range provides a test for the charge inde-pendence of the nuclear force by a comparison of the experimental value with the theoretical predictions by NOYES /4.27/ (2.73 fm) and BREIT et al. /4.28/ (2.83 fm) which were derived from pp scattering under the assumption of charge independence. An estimate of the singlet scattering length (a_s) under the same assumption, how-ever, leads to a value of -17 fm, which is quite different from the experimental results (-23.7 fm).

4.5.2 Experiments and Results

The (n,p) scattering length b_p could be determined exactly only by transmission ex-periments on para-hydrogen (SQUIRES et al. /4.29/ and CALLERAME et al. /4.30/) and by mirror reflection measurements on hydrogen containing liquids (BURGY et al. /3.26/, DICKINSON et al. /3.29/ and KOESTER et al. /3.38/). The most reliable data given in

Table 4.4. Experimental scattering lengths and free cross sections of the neutron-proton interaction

Authors Year	Method	b_p (fm)	Relative standard deviation %	Deviation %
SHULL 1951	NaH Diffr.	-4.0(2)	5	-6.5
SQUIRES 1955	Para-Hydr.	-3.80(5)	1.3	-1.6
CALLERAME 1975	Para-Hydr.	-3.733(4)	0.1	+0.2
BURGY 1951	Mirror-Refl.	-3.78(2)	0.6	-1
DICKINSON 1962	Mirror-Refl.	-3.74(2)	0.6	+0.02
KOESTER 1975	Gravity-Refr.	-3.7409(11)	0.03	0
		σ_0 (b)		
MELKONIAN 1949	TOF[a]	20.36(10)	0.5	-0.7
NEILL 1968	TOF	20.366(76)	0.4	-0.6
HOUK 1971	TOF	20.436(23)	0.1	-0.3
DILG 1975	Res.Scatt.[b]	20.491(14)	0.07	0

a TOF = time of flight measurement.

b Double resonance scattering.

Table 4.4 show for both methods a drastic improvement of the accuracy by one order of magnitude in the most recent experiments. The two most accurate values were obtained by different methods, however, they are in satisfactory agreement. Furthermore, the deviations from the earlier results are of the order of their standard derivations. It should be mentioned that the extremently high accuracy of the gravity-refractometer experiment became possible by measurements on 18 various liquids of 11 different organic substances containing carbon, hydrogen and/or chlorine. The purity of the compounds was carefully estimated.

Considering the values for the free cross section σ_0 per proton (hydrogen atom) in Table 4.4, a likely high improvement of the accuracy in recent years can be established as to scattering lengths, but there exists little agreement between the most exact values. All results are smaller than the most precise value reported by DILG /3.52/. The discussion of this discrepancy by Dilg, however, supported his higher value.

From the most precise numbers at present

$$b_p = -3.7409(11) \text{ fm and } \sigma_o = 20.491(14) \text{ b}$$

the low-energy (n,p) parameters are found to be /3.52, 38/

$$a_t = 5.424 \pm 0.003 \text{ fm and } a_s = -23.749 \pm 0.008 \text{ fm.}$$

Moreover, with the deuteron radius

$$R = 4.31767(9) \text{ fm}$$

/3.51/ and with five single-energy data for $\sigma(k)$ /3.52/ the further parameters are as follows:
the mixed triplet effective range

$$\rho_{ot}(-\varepsilon,o) = 2R(1-R/a_t) = 1.760(5) \text{ fm}$$

and the singlet range

$$r_{os} = 2.77(5) \text{ fm}$$

in the shape independent approximation. The determination of $r_{ot} = \rho(0,0)$ from the mixed range $\rho(-\varepsilon,0)$ is model dependent /4.24/. Applying the shape correction proposed by NOYES /4.31/ one obtains the effective range to be $r_{ot} = 1.747(6)$ fm.
  The new $r_{os} = 2.77(5)$ fm as well as the value given by LOMON and WILSON /4.26/ $r_{os} = 2.76(5)$ lie in the range of the theoretical prediction based on charge independence, but we must remember that the measured $a_s = -23.749(8)$ fm differs appreciably from the predicted -17 fm calculated also on the basis of charge independence.

4.6 Neutron-Deuteron Interaction

4.6.1 Introduction

The study of the low-energy neutron-deuteron interaction is of fundamental interest to the problem of nuclear force for the following reasons: Free of electromagnetic interactions, the neutron-deuteron (n,d) scattering may contribute directly to the problem of existence and strength of any explicit three-nucleon force, or, on the other hand, the accuracy of the existing potential models for the two-nucleon interaction can be rigorously tested because the behaviour of the (n,d) system is much more sensitive than the deuteron to the details of the nuclear force.

The low energy (n,d) scattering parameter are calculable with sufficient precision, for instance, using simple two-nucleon interactions. Comparison with the results of calculations for the bound state, e.g. the triton binding energy, then became meaningful.

4.6.2 Experimental Results

Of particular importance for the three-nucleon study is the zero-energy (n,d) scattering, which is completely parametrized by the doublet (^2a) and the quarted (^4a) scattering length. The former is of special interest, since it provides, together with the triton binding energy, the above mentioned sensitive test of nuclear three-body calculations. The two parameters ^4a, ^2a of the scattering by the free deuteron are related to the bound coherent scattering lengths b_D, to the zero-energy free scattering scros section σ_o or to the bound incoherent cross section $\sigma_{inc,b}$ as follows

$$b_D = {}^4a + (1/2)^2a; \quad \sigma_o = (4\pi/3)(2 \cdot {}^4a^2 + {}^2a^2) \text{ and } \sigma_{inc,b} = 2\pi({}^4a - {}^2a)^2.$$

Experimental determinations of b_D were performed by BARTOLINI et al. /3.30/ using mirror reflexion from D_2O and by KOESTER et al. /3.43/, DILG et al. /3.53/ and NIST-LER /4.32/ who all employed the advanced technique of the Christiansen-filter and the gravity refractometer. The free cross section was exactly measured only by FERMI et al. /3.57/ utilizing the resonance detectors In (1.44 eV) and Ag (5.2 eV) and by DILG et al. /3.53/ with 132 eV neutrons from Co- resonance scattering. GISS-LER /4.33/ reported a value of σ_{inc}, which was obtained by transmission measurements with D_2O at temperatures of 4 K and 77 K and at neutron energies in the range from $3 \cdot 10^{-5}$ eV to $3 \cdot 10^{-3}$ eV. Further information on the scattering parameters was derived from the ortho- and para-deuterium cross sections by NIKITIN et al. /4.34/ and most recently by CALLERAME et al. /4.35/ as well as from the angular variation of slow neutron scattering from deuterium gas by HURST et al. /4.36/. The ratio $^4a/^2a > 1$ was established by ALKIMENKOV et al. /4.37/ using polarized neutrons and deuterons in a transmission experiment.

The experimental situation, as demonstrated in Table 4.5 and Fig.4.1, looks like a puzzle without any solution. Nevertheless, four sets of scattering parameters (^4a and ^2a) could be deduced. A good accordance exists between the three sets A, B and D, whereas set C lies far outside. The great deviation of C is mainly due to the value of the coherent scattering length $b_D = 6.21(4)$ fm obtained by BARTOLINI et al. /3.30/ in 1968, by means of liquid mirror reflection from D_2O. This b_D is also in sharp disagreement with the values given by KOESTER /3.43/ and most recently by NISTLER /4.32/. The latter high precision value and the value given by DILG et

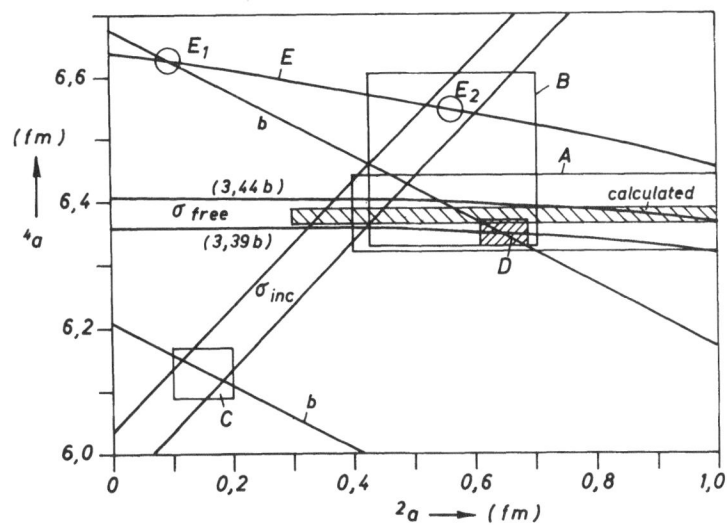

<u>Fig.4.1.</u> (n,d) singlet and quartet scattering lengths. Comparison of the results of various experiments and of calculations

al. /3.53/ were also obtained by mirror reflection from D_2O. On the other hand, the Livermore group (BARTOLINI et al.) and the Munich group (KOESTER, DILG, and NISTLER) both reported results of reflection measurements on Hg- and F-containing mirrors which are in close agreement within the limit of error. Thus, the observed discrepancy in the D_2O results might be due to a faulty determination of the D- and 0^{18}-content of the mirror substances. NISTLER /4.32/, therefore, investigated the isotopic composition of his mirrors by very precise measurement of the density and by nuclear-resonance spectroscopy. In addition, he performed measurements on 7 various mirrors consisting of different D_2O-H_2O mixtures. All experiments resulted in b_D = 6.674 $\pm$ 0.002 fm in constrast to the result of the Livermore group. Furthermore, the "Munich" set D implies the value σ_o = 3.390 $\pm$ 0.012 b for the free cross section, which was recently confirmed by STOLER et al. /4.38/ by transmission measurements of the neutron total cross section of deuterium from 1 to 1000 keV. These measurements led also to an extrapolated σ_o = 3.396 at zero energy.

Consequently, we may suggest in accordance with LEVINGER /4.39/

set D: ^4a = 6.35 $\pm$ 0.02 fm and ^2a = 0.65 $\pm$ 0.04 fm

for future use. In order to be able to do so, we must ignore the result of transmission experiments on ortho-deuterium recently given by CALLERAME et al. /4.35/.

Table 4.5. Slow neutron scattering parameters of the (nd,) interaction. The measured values are underlined

Author Year	Cross sections (b)		Scattering lengths (fm)		
	σ_0	σ_{inc}	b	^2a	^4a
HURST et al. 1951			$\underline{^2a/^4a = 0.12(4)}$		
FERMI et al. 1949	$\underline{3.44(6)}$				
derived (A)		2.03(13)	6.73(18)	0.7(3)	6.38(6)
NIKITIN et al. 1955 (B)				0.57(14)	6.47(14)
GISSLER 1963		$\underline{2.25(4)}$			
BARTOLINI et al. 1968			$\underline{6.21(4)}$		
derived (C)	3.15(4)			0.15(5)	6.13(4)
DILG et al. 1971	$\underline{3.390(12)}$		$\underline{6.672(7)}$		
KOESTER et al. 1968			$\underline{6.70(5)}$		
NISTLER 1974			$\underline{6.674(2)}$		
derived (D)		2.04(3)		0.65(4)	6.35(2)
CALLERAME 1975	$\underline{(2^4a+^2a)^2 + (5/4)(^4a-^2a)^2 = 2.313(3) \ b}$				
derived (E_1)	3.68	2.68	$\underline{6.674(2)}$	0.10	6.625
(E_2)	3.61	$\underline{2.25}$	6.84(3)	0.57(4)	6.55(2)

This result combined with another parameter leads to some very different sets E, which are incompatible with the other sets with the possible exception of set B.

4.6.3 Theoretical Calculations

There are numerous calculations of the deuteron scattering parameters and of the triton binding energy mainly based on two nucleon interactions. Excellent reviews of the nuclear two- and three-body problems are given by LEVINGER /4.39/ (1974) and, with the status of 1968, by DELVES and PHILLIPS /4.40/. References about the theoretical works can be found there. The discussion sofar was dominated by the discrepancy between the two sets A and C. It was shown that all more exact calculations fail to fit simultaneously the binding energy and the doublet scattering length of set C and favour set A. The numerical results for ^2a fall on or close to the so-called Phillips line for which HARMS /4.41/ gave the equation $^2a = 0.75 (E_T + 8.5) + 0.75$ fm. If E_T is near the experimental -8.48 MeV, then ^2a will also be near the experimental value of set D. Results of some theoretical works in the last decade are compiled in Table 4.6. One sees that the recent calculations (lines 1 to 3) yield numerical results which agree closely with the experimental values for the triton binding energy and with the scattering parameters of set D, see, e.g., TJON /4.44/.

Table 4.6. Calculated parameters of the three body system (n+d)

Author, year (potential shape, method)	$-E_T$(MeV)[a]	^2a(fm)	^4a(fm)	Comments
ALT, BAKKER 1975 /4.42/ separable, local soft core	10.165	-0.411	6.386	
	9.439	-0.2341		
	8.643	0.851		
	8.630	0.936	6.394	
BELYAEV 1971 /4.43/ "relativistic potential"	8.10	1.33	-	
	8.56	1.15	6.37	
	9.12	0.54	6.35	
	$-E_T$(MeV)	^2a(fm)	r_{os}(fm)	
TJON 1970 /4.44/ central, local Yukawa potentials Faddeev equation	9.1	-0.4	2.6	no core
	8.4	0.3	2.8	no core
	8.8	0.4	2.6	with core
	8.3	0.9	2.8	with core
DELVES et al. 1969 /4.45/ local, H.J.-potential (n,p) variational calculation	7	1.2 ± 1.0		
	8.4	0.8 ± 1.0		
PHILLIPS 1968 /4.46/ separable, tensor component	8.6	+0.75	2.7	5.5 %[b]
	9.2	+0.33	2.7	5 %
	8.05	1.12	2.7	7 %

a The experimental values of E_T is - 8.48 MeV.

b Deuterium D-state contribution.

Reviewing both the experimental and theoretical situation, the results are summarized in Fig.4.1 in which it is shown how the area of sets A, B and D coincide with the area including the theoretical results. Sets C, E_1 and E_2 however, lie in part far outside the theoretical predictions so that they must be considered as less probable at present.

The result reported by Callerame and given as line E in the figure exhibits the greatest problem. To clarify experimentally the discrepancies a repetition of the ortho-deuterium experiment as well as a new direct determination of σ_{inc} may be useful.

References

1.1 R.P. Ozerov, I.D. Datt: Atomic Energy Review 13, 651 (1971)
2.1 E.R. Cohen, B.N. Taylor: J. Phys. Chem. Ref. Data 2, 663 (1973)
2.2 H.A. Bethe, P. Morrison: Elementary Nuclear Theory, 2nd edition, (John Wiley, New York, 1956)
2.3 J.M. Blatt, J.D. Jackson: Phys. Rev. 76, 18 (1949); H.A. Bethe, ibid 76, 38 (1949)
2.4 H. Feshbach, C.E. Porter, V.F. Weisskopf: Phys. Rev. 96, 448 (1954)
2.5 K.K. Seth, D.J. Hughes, R.L. Zimmermann, R.C. Garth: Phys. Rev. 110, 692 (1958)
2.6 F.W.K. Firk, J.E. Lynn, M.C. Moxon: Proc. Phys. Soc. London 82, 477 (1963)
2.7 J. Morgenstern, G. Bianchi, C. Corge, V.-D. Huynh, J. Julien, F. Netter, G. Le Pottevin, M. Vaster; Nucl. Phys. 62, 529 (1965)
2.8 S.W. Lovesey: J. Phys. C (Solid State Phys.) 2, 981 (1969)
2.9 J. Schwinger: Phys. Rev. 73, 407 (1948)
2.10 L.L. Foldy: Phys. Rev. 83, 688 (1955); Rev. Modern Phys. 30, 471 (1958)
2.11 K. Binder: Report PTHM-FRM 101, Physik-Department der Technischen Universität München (1970)
2.12 L. Koester, W. Nistler, W. Waschkowski: Phys. Rev. Letters 36, 1021 (1976)
2.13 I.I. Gurevich, L.V. Tarasov: Low Energy Neutron Physics (North-Holland Publ. Co., Amsterdam 1968)
2.14 H. Eckstein: Phys. Rev. 89, 490 (1953)
2.15 M.L. Goldberger, F. Seitz: Phys. Rev. 71, 294 (1947)
2.16 K. Binder: Phys. Stat. Sol. 41, 767 (1970)
2.17 G. Placzek: Phys. Rev. 86, 377 (1952)
2.18 G. Placzek: Phys. Rev. 93, 895 (1954)
2.19 G. Placzek: Phys. Rev. 105, 1240 (1957)
2.20 G. Placzek, L. van Hove: Nuovo Cimento 1, 233 (1955)
2.21 G. Placzek, B.R.A. Nijboer, L. van Hove: Phys. Rev. 82, 392 (1951)
2.22 G.C. Wick: Phys. Rev. 94, 1228 (1954)
2.23 L. Koester, K. Knopf, W. Waschkowski: Z. Physik A277, 77 (1976)
3.1 E.O. Wollan, C.G. Shull: Phys. Rev. 73, 822 (1948)
3.2 C.G. Shull, E.O. Wollan: Phys. Rev. 81, 527 (1951)
3.3 M.K. Wilkinson, E.O. Wollan, W.C. Koehler: Ann. Rev. Nucl. Sci. 11, 304 (1961)
3.4 H. Maier-Leibnitz, T. Springer: Ann. Rev. Nucl. Sci. 16, 207 (1966)
3.5 G.E. Bacon: Neutron Diffraction, 3rd edition (Clarendon Press, Oxford, Great Britain 1975)
3.6 B.T.E. Willis: Chemical Application of Thermal Neutron Scattering (Oxford University Press 1973)
3.7 B.W. Batterman, H. Cole: Rev. Mod. Phys. 36, 681 (1964)
3.8 N. Kato: Acta Cryst. 14, 526 (1961) and 14, 627 (1961)
3.9 C.G. Shull, W.M. Shaw: Z. Naturforsch. 28a, 657 (1973)
3.10 C.G. Shull: Phys. Rev. Lett. 21, 1585 (1968)

3.11 C.G. Shull: J. Appl. Cryst. 6, 257 (1973)
3.12 H. Maier-Leibnitz, T. Springer: Z. Physik 167, 386 (1962)
3.13 C.G. Shull, K.W. Billmann, F.A. Wedgwood: Phys. Rev. 153, 1415 (1967)
3.14 C.S. Schneider, C.G. Shull: Phys. Rev. B3, 830 (1970)
3.15 C.G. Shull: Refractive Index Studies with Neutrons in: Some Lectures on Neu-
      tron Physics, Summer School Alushta, May 1969, Dubna report 3-4981, USSR
3.16 F.D. Landkammer: Z. Physik 189, 113 (1966)
3.17 H. Friedrich, W. Heintz, F. Piefke: Physikalisch-Technische Bundesanstalt,
      Braunschweig, GFR, Report PTB-FMRB-59 (1974) and private communication (1976)
3.18 U. Bonse, M. Hart: Appl. Phys. Lett. 6, 155 (1965)
3.19 U. Bonse, M. Hart: Z. Physik 194, 1 (1966)
3.20 H. Rauch: Acta Phys. Austr. 33, 50 (1971)
3.21 C.G. Shull: J. Appl. Cryst. 6, 257 (1973)
3.22 H. Rauch, W. Treimer, U. Bonse: Phys. Lett. 47A, 369 (1974)
      H. Rauch, M. Suda: Phys. Stat. Sol. (a) 25, 495 (1974)
3.23 W. Bauspiess, U. Bonse, H. Rauch, W. Treimer: Z. Physik 271, 177 (1974)
3.24 H. Rauch et al: 1976 Int. Conf. on the Interactions of Neutrons with Nuclei,
      Session PD2 July 6-9, 1976, Lowell, Mass., USA
3.25 P. Korpiun: Z. Physik 195, 146 (1966)
3.26 M.T. Burgy, G.R. Ringo, D.J. Hughes: Phys. Rev. 84, 1160 (1951)
3.27 E. Fermi, L. Marshall: Phys. Rev. 71, 666 (1947)
3.28 D.J. Hughes, M.T. Burgy: Phys. Rev. 81, 498 (1951)
3.29 W.C. Dickinson, L. Passel, O. Halpern: Phys. Rev. 126, 632 (1962)
3.30 W. Bartolini, R.E. Donaldson, D.J. Groves: Phys. Rev. 174, 313 (1968)
3.31 C.J. Heindl, I.W. Rudermann, J.M. Ostrowski, J.R. Ligenza, D.M. Gardner: Rev.
      Sci. Instr. 27, 620 (1956)
3.32 D. Bally, S. Todireanu, S. Ripeanu, M.G. Belloni: Rev. Sci. Instr. 33, 916
      (1962)
3.33 A.W. McReynolds: Phys. Rev. 84, 969 (1951)
3.34 M.F. Crouch, V.E. Krohn, G.R. Ringo: Phys. Rev. 102, 1321 (1956)
3.35 H. Maier-Leibnitz: Z. angew. Physik 14, 738 (1962)
3.36 L. Koester: Z. Physik 182, 328 (1965)
3.37 L. Koester: Z. Physik 198, 187 (1967)
3.38 L. Koester, W. Nistler: Z. Physik A272, 189 (1975)
3.39 W.-D. Trüstedt: Z. Naturforsch. 26a, 400 (1971)
3.40 N. Nücker: Z. Physik 227, 152 (1969)
3.41 C. Christiansen: Wied. Annalen 23, 298 (1884)
3.42 R.J. Weiss: Phys. Rev. 83, 379 (1951)
3.43 L. Koester, H. Ungerer: Z. Physik 219, 300 (1969)
3.44 L, Koester, K. Knopf: Z. Naturforsch. 26a, 391 (1971)
3.45 E. Wunderlich: Diplomarbeit Fachbereich Physik, Technische Universität Mün-
      chen (1975), Reaktorstation Garching
3.46 J. Meier: Fachbereich Physik, Technische Universität München, work in pro-
      gress (1976)
3.47 R.E. Donaldson, W. Bartolini, H. Otsuki: Phys. Rev. C5, 1952 (1972)
3.48 R.E. Donaldson, D.J. Groves, R.K. Pearson: Phys. Rev. 146, 660 (1966)
3.49 L. Koester, K. Knopf: Z. Naturforsch. 27a, 901 (1972)
3.50 S.F. Mughabghab, D.J. Garber: Brookhaven National Laboratory Report BNL 325,
      3rd edition 1974, Vol. 1
3.51 T.L. Houk: Phys. Rev. C3, 1886 (1971)
3.52 W. Dilg: Phys. Rev. C11, 103 (1975)
3.53 W. Dilg, L. Koester, W. Nistler: Phys. Lett. 36B, 208 (1971)
3.54 W. Dilg, H. Vonach: Z. Naturforsch. 26a, 442 (1971)
3.55 W. Dilg: Nucl. Instr. Meth. 122, 343 (1974)
3.56 D.G. Hurst, J. Alcock: Canad. J. Phys. 29, 36 (1951)
3.57 E. Fermi, L. Marshall: Phys. Rev. 75, 578 (1949)
3.58 L.J. Rainwater, W.W. Havens: Phys. Rev. 70, 136 (1946)
3.59 B.D. McDaniel: Phys. Rev. 70, 832 (1946)
3.60 E. Melkonian: Phys. Rev. 76, 1744 (1949)
3.61 C.T. Hibdon, C.O. Muehlhouse: Phys. Rev. 76, 100 (1949)

3.62 L.A. Rayburn, E.O. Wollan: Nucl. Phys. 61, 381 (1965)
3.63 L. Koester, P. Schacht: Z. angew. Physik 31, 21 (1971)
3.64 W. Dilg, H. Vonach: Nucl. Instr. Meth. 100, 83 (1972)
3.65 W. Waschkowski, L. Koester: Z. Naturforsch. 31a, 115 (1976)
3.66 C.S. Schneider: Acta Cryst. A32, 375 (1976)
4.1   B.T.E. Willis: Chemical Applications of Thermal Neutron Scattering (Oxford
      University Press (1973)
4.2   W. Schmatz, T. Springer, J. Schelten, K. Ibel: J. Appl. Cryst. 7, 96 (1974)
4.3   J. Estermann, O.C. Simpson, O. Stern: Phys. Rev. 71, 238 (1947)
4.4   A.W. McReynolds: Phys. Rev. 83, 172, 233 (1951)
4.5   J.W.T. Dabbs, J.A. Harvey, D. Paya, H. Horstmann: Phys. Rev. 139, B756 (1965)
4.6   L. Koester: Z. Physik 198, 187 (1967)
4.7   F.C. Witteborn, W.M. Fairbank: Phys. Rev. Lett. 19, 1049 (1967)
4.8   R.V. Pound, J.L. Snider: Phys. Rev. 140, B788 (1965)
4.9   L. Koester: Phys. Rev. D14, 907 (1976)
4.10 A.W. McReynolds: Bull. Amer. Phys. Soc. 12, 105 (1967)
4.11 R. Collela, A.W. Overhauser, S.A. Werner: Phys. Rev. Lett. 23, 1472 (1975)
4.12 R. v. Eötvos, D. Pekar, E. Fekete: Ann. Physik 68, 11 (1922)
4.13 P.G. Roll, R. Krotkov, R.H. Dicke: Ann. Phys. (N.Y.) 26, 442 (1964)
4.14 J.G. Williams et al.: Phys. Rev. Lett. 36, 551 (1976)
4.15 J.J. Shapiro, C.C. Counselmann, III., R.W. King: Phys. Rev. Lett. 36, 555 (1976)
4.16 E. Fermi, L. Marshall: Phys. Rev. 72, 1139 (1947)
4.17 V.E. Krohn, G.R. Ringo: Phys. Rev. 148, 1303 (1966) ibid, D8, 1305 (1973)
4.18 W.W. Havens, Jr., I.I. Rabi, J.L. Rainwater: Phys. Rev. 72, 634 (1947)
4.19 E. Melkonian, B.M. Rustad, W.W. Havens, Jr.: Phys. Rev. 114, 1571 (1959)
4.20 D.J. Hughes, J.A. Harvey, M.D. Goldberg, M.J. Stafne: Phys. Rev. 90, 497 (1953)
4.21 Yu.A. Alexandrov: Communications of the Joint Institute for Nuclear Research,
      Dubna, USSR, E3-5713 (1971) (in English)
4.22 Yu.A. Alexandrov, T.A. Machekhina, L.N. Sedlakova, L.E. Fykin: Preprint.
      Joint Institute for Nuclear Research Dubna, USSR, P3-7745 (1974)
4.23 M.J. Moravcsik: Rept. Progr. Phys. 35, 587 (1972)
4.24 H.P. Noyes: Ann. Rev. Nucl. Sci. 22, 465 (1972)
4.25 H.P. Noyes: Phys. Rev. 130, 2025 (1963)
4.26 E. Lomon, R. Wilson: Phys. Rev. C9, 1329 (1974)
4.27 H.P. Noyes, H.M. Lipinski: Phys. Rev. C4, 995 (1971)
4.28 G. Breit, K.A. Friedmann, J.M. Holt, R.E. Seamon: Phys. Rev. 170, 1424 (1968)
4.29 G.L. Squires, A.T. Stewart: Proc. Roy. Soc. (London) A230, 19 (1955)
4.30 J. Callerame, D.J. Larson, S.J. Lipson, R. Wilson: Phys. Rev. C12, 1432 (1975)
4.31 H.P. Noyes: Nucl. Phys. 74, 508 (1965)
4.32 W. Nistler: Z. Naturforsch. 29a, 1284 (1974)
4.33 W. Gissler: Z. Kristallographie 118, 149 (1963)
4.34 S.J. Nikitin et al.: First Geneva Conf., Vol. 2, 81 (1955)
4.35 J. Callerame, D.J. Larson, S.J. Lipson, R. Wilson: Phys. Rev. C12, 1428 (1975)
4.36 C.D. Hurst, J.Alcock: Canad. J. Phys. 29, 36 (1951)
4.37 V.P. Alkimenkov et al.: Phys. Lett. 24B, 151 (1967)
4.38 P. Stoler, N.N. Kauskal, F. Green, E. Harms, L. Laroze: Phys. Rev. Lett. 29,
      1745 (1972)
4.39 J.S. Levinger: Springer Tracts in Modern Physics 71, 88 (1974)
4.40 L.M. Delves, A.C. Phillips: Rev. Mod. Phys. 41, 497 (1969)
4.41 E. Harms: XIV Latin American School of Physics, Caracas, July 1972 (Reidel,
      Holland)
4.42 E.O. Alt, B.L.G. Bakker: Z. Physik A273, 37 (1975)
4.43 V.B. Belyaev, E. Wrzecianko, A.L. Zubarev: Soviet J. Nucl. Phys. 12, 503 (1971)
4.44 J.A. Tjon: Phys. Rev. D1, 2109 (1970)
4.45 L.M. Delves, J.M. Blatt, I. Pask, B. Davies: Phys. Lett. 28B, 472 (1969)
4.46 A.C. Phillips: Nucl. Phys. A107, 209 (1968)
4.47 A. Pais: Rev. Mod. Phys. 38, 215 (1966)
4.48 A. Le Yaouanc, L. Oliver, O. Pene, J.-C. Raynal: Phys. Rev. D12, 2137 (1975)

4.49 W. Schmatz, B. Bauer, M. Löwenhaupt: Nuclear Cross Sections and Technology,
NBS SP 425, 823 (1975) Be, Al, Si, Nb, Nd;
W. Schmatz, private communication (1976) Mg, Ti, Fe, Cu, Zn, Y, Pd, Ag, Sn,
La, Ta, W, Pt;
R. Scherm, Nukleonik 12, 4 (1968) Pb, Bi;
4.50 A. Abragam et al.: Phys. Rev. Lett. 28, 305 (1972) F;
P. Roubeau, A. Abragam, G.L. Bacchella, H. Glättli, A. Malinovki, P. Meriel,
J. Piesveaux, M. Pinot: Phys. Rev. Lett. 33, 102 (1974) Al, Sc, Zr, Nb, Ta;
A. Abragam, G.L. Bacchella, H. Glättli, P. Meriel, J. Piesveaux, M. Pinot:
J. Physique 36, L-263 (1975) Na, Co, Zr, Pt, Au.

Very Low Energy Neutrons

Albert Steyerl

1. Introduction

In the past decade investigators have begun to explore the special features and potentialities of neutrons in the region of energies lying significantly below the
energy range of "cold neutrons", i.e. substantially below the thermal equilibrium
energy of $\approx 10^{-3}$-10^{-2} eV of neutron spectra from low-temperature moderators ("cold
sources") in primary neutron sources. The research with "even colder" neutrons may
be classified into three branches according to the main objectives pursued:

a) Investigation of the interaction of very slow neutrons with matter and utilization of their special scattering features for the study of submicroscopic structures
and low-energy excitations within and at the surface of solids. This work was ini-
tiated by Maier-Leibnitz in Munich in 1966 /1.1/.

b) Study and application of a peculiar property of neutrons with energies $E \lesssim$
10^{-7} eV: According to the well-known dispersion law of neutron optics /1.2/ neutrons
with sufficiently low energy undergo total external reflection from many materials
at any angle of incidence. This unique phenomenon is the basis for neutron containment in closed material cavities as proposed first by ZEL'DOVICH in 1959 /1.3/ and
pioneered by a group of the late F.L. Shapiro in Dubna in 1968 /1.4/. The most promising future application of neutron storage in material "neutron bottles" is a
high-sensitivity experiment to search for an electric dipole moment (EDM) of the
neutron, as proposed first by SHAPIRO in 1968 /1.5/. Such an experiment is presently in preparation in Leningrad, Dubna, and at Sussex University. Moreover, it has
been proposed by KASHUKEEV in 1971 /1.6/ to use stored neutrons for a sensitive
search for an electrical charge of the neutron.

c) Investigation of the interaction of very low energy neutrons with magnetic
fields and, in particular, following proposals of PAUL /1.7/ and VLADIMIRSKY /1.8/,
development of a magnetic storage ring for neutrons of energy $E \approx 10^{-6}$ eV, with
the ultimate aim of a precise measurement of the neutron's lifetime for β-decay.
Such work is now in progress in Bonn.

This synopsis shows that perhaps the most interesting results of work with very
low energy neutrons are to be expected only in the years to come. In this sense a
survey of the new field at this point may be considered a preview rather than a review. The main objective of this article is to report on the intensive research

carried out in a number of laboratories, since SHAPIRO's first progress report of this kind in 1972 /1.9/, into the development of the special experimental techniques required for the efficient production, channeling, spectroscopy, storage, polarization, magnetic resonance and detection of very low energy neutrons. Attention will also be paid to the special features of their interaction with matter, in particular in connection with the surprisingly short experimental lifetimes in material bottles which have stimulated a number of theoretical investigations into the possible reasons for the apparent anomaly observed.

A remark should be made on the nomenclature used for very low energy neutrons. Originally, neutrons with energies $E < 10^{-4}$ eV (1 K) were defined to be "ultracold neutrons" (UCN) by GUREVICH and TARASOV /1.10/. In the meantime, however, it has been proposed by FRANK /1.11/ to restrict the terminology "ultracold" only to neutrons capable of undergoing total external reflection even at normal incidence in a given material or magnetic device (or in a part of it), and to designate neutrons with energies $E < 10^{-4}$ eV, but exceeding the respective critical value, as "very cold neutrons" (VCN). This nomenclature will generally be used in the present review, although not in a strict sense, since there are cases where a rigorous distinction is obviously meaningless, e.g. where neutrons are accelerated over the threshold, or where the reflection properties are irrelevant.

In this article we present first a brief outline of some general aspects of very-low-energy neutron physics. In Section 3 the various techniques and problems of VCN production, guidance, and detection will be discussed. Section 4 deals with VCN cross-section measurements and their implications. Section 5 gives an outline of UCN storage experiments and of the various attempts to explain the experimental results. Recent progress made in preparation for the specific experiments of a search for the neutrons's electric dipole moment using UCN, and of constructing a magnetic bottle for neutrons will be discussed in Section 6.

2. Fundamentals of Very Low Energy Neutrons

In the first place it seems worthwhile to emphasize that we cannot expect very slow neutrons to show any peculiar property that would be fundamentally different from the characteristics of faster neutrons. In theory, a fast neutron may readily be changed into an arbitrarily slow one, or vice versa, by the application of a suitable Lorentz transformation. All the known neutron interactions remain unchanged by this process inasmuch as Lorentz invariance is strictly satisfied. However, at very low energies the strong and electromagnetic interactions give rise to a number of interesting new phenomena and attractive applications which seem to justify the efforts made for UCN production (which in practice is considerably more troublesome than the application of a Lorentz transformation!) and their utilization.

2.1 Index of Refraction

As in optics, the concept of the refractive index plays a fundamental part in the description of the interaction of very low energy neutrons with matter and magnetic fields. It is well known that refraction of waves in dense systems is microscopically described as being due to coherent elastic scattering in the forward direction. The physical phenomena involved are simpler for neutrons than, say, for photons, because slow neutrons are scattered isotropically. In order to familiarize the reader with some important concepts, we give an elementary but quite general derivation of the refractive index for neutrons and its dispersion law in the case of nonmagnetic scattering (following FOLDY /2.1/).

Consider a neutron wave $\Psi_0(\vec{r})$ satisfying the wave equation in vacuum

$$(\nabla^2 + k^2)\Psi_0(\vec{r}) = 0, \tag{2.1}$$

where k is a wavenumber characterizing Ψ_0. Let Ψ_0 impinge on an arbitrarily shaped region in space which is densely filled with atoms fixed at the points $\vec{r}_j$. Then the total wave field $\Psi(\vec{r})$ is given as the superposition of the incident wave $\Psi_0(\vec{r})$ and all the spherical scattering waves $-b_j\Psi(\vec{r}_j) \dfrac{\exp(ik|\vec{r} - \vec{r}_j|)}{|\vec{r} - \vec{r}_j|}$ originating from any nucleus j as a result of scattering of the wave incident on it, which in all cases of interest here may be identified with the locally averaged wave $\Psi(\vec{r}_j)$ [1]. b_j denotes the bound-atom coherent-scattering length of nucleus j (or its average value, if there is an isotopic or spin disorder). For long-wavelength neutrons the summation over the scattering centres may always be replaced by an integration. Thus $\Psi(\vec{r})$ is determined by the integral equation

$$\Psi(\vec{r}) = \Psi_0(\vec{r}) - \int \frac{e^{ik|\vec{r}-\vec{r}'|}}{|\vec{r} - \vec{r}'|} \, \rho(\vec{r}')\Psi(\vec{r}')d\vec{r}'. \tag{2.2}$$

$\rho(\vec{r})$ is the "scattering-length density" defined by

$$\rho(\vec{r}) = \frac{1}{V} \sum_{\vec{r}_j \text{ in } V} b_j, \tag{2.3}$$

---

[1] Strictly speaking, the "effective field" acting on the nonisolated nucleus j is slightly different from $\Psi(\vec{r}_j)$, as pointed out first by LAX /2.2, 3/ and EKSTEIN /2.4, 5/, the difference depending on the configuration of scatterers and increasing with packing density. However, the calculations of IGNATOVICH /2.6/, IGNATOVICH and LUSHCHIKOV /2.7/, and of LENK /2.8/ show that for UCN the resulting corrections are negligible both in crystalline and in disordered media.

where the summation extends over the atoms lying in a macroscopically small volume element V at $\vec{r}$, yet large enough to contain many atoms. For a homogeneous medium with N atoms with scattering length b in unit volume: $\rho(\vec{r})$ = Nb inside the medium, and = 0 outside.

The integral equation (2.2) is equivalent to the differential equation

$$\left[\nabla^2 + K^2(\vec{r})\right]\Psi(\vec{r}) = 0,\tag{2.4}$$

where

$$K(\vec{r}) = \sqrt{k^2 - 4\pi\rho(\vec{r})}\tag{2.5}$$

is the local wavenumber. $K(\vec{r})$ may be expressed through a local <u>index of refraction</u>

$$n(\vec{r}) = K(\vec{r})/k = \sqrt{1 - 4\pi\rho(\vec{r})/k^2}.\tag{2.6}$$

The usual continuity conditions at boundaries and the specification of the incident wave are incorporated in the integral equation (2.2), which thus fully determines the reflection, refraction and scattering properties of the medium. (This last statement applies in a rigorous sense only to the case where any incoherent effects are absent, including the possibility of inelastic or spin-flip scattering, or nuclear capture).

The dispersion law (2.6) is of a very special form in the sense that exactly the same relation would hold if the medium were thought to be replaced by a potential

$$U(\vec{r}) = \frac{2\pi\hbar^2}{m}\rho(\vec{r}),\tag{2.7}$$

where m is the neutron mass. $U(\vec{r})$ is the local space-average of the Fermi pseudo-potential, $(2\pi\hbar^2/m)b_j\delta(\vec{r}-\vec{r}_j)$, which describes the slow-neutron coherent-scattering interaction with nucleus j. In contrast with the usual situation in optics, the scattering potential for slow neutrons is independent of wavelength over a wide range of low energies (say, E < 1 keV).

In the presence of a magnetic induction field $\vec{B}(\vec{r})$ there is an additional interaction energy $-\vec{\mu}\cdot\vec{B}$ with the neutron magnetic moment $\vec{\mu}$. Hence

$$U(\vec{r}) = \frac{2\pi\hbar^2}{m}\rho(\vec{r}) \pm \mu B(\vec{r}),\tag{2.8}$$

where the positive and negative sign holds respectively for neutron spin configuration parallel and antiparallel to $\vec{B}$.

For most nuclei the coherent scattering length is positive, hence the refractive index is less than 1 and the scattering potential is repulsive. U reaches a few times

10^{-7} eV for selected substances as well as for high magnetic fields. A few typical values are given in Table 2.1 which also shows the effect of saturation magnetization (at room temperature) in a few ferromagnets.

Table 2.1. Values of the scattering potential for various materials are shown with the corresponding critical velocities v_{cr} for total reflection at normal incidence

Substance	$U \left[10^{-7} eV \right]$ [a]	$v_{cr} \left[m/s \right]$
Al	0.541	3.22
Mg	0.595	3.37
Cu	1.65	5.6
C (graphite, density 2 g/cm^3)	1.75	5.8
Be	2.5	6.9
BeO (2.9 g/cm^3)	2.5	6.9
D_2O (1.105 g/cm^3)	1.657	5.63
Fe	3.50	8.2
	0.90	4.15
Ni	2.7	7.2
	2.0	6.2
Co	1.74	5.76
	-0.42	-
Mn	-0.68	-
Ti	-0.50	-
H_2 (liquid 20 K)	-0.081	-
H_2O (1.00 g/cm^3)	-0.146	-
Polythene	-0.07	-

[a] For ferromagnets the values $U_{\pm} = U_{nuclear} \pm \mu B$ are given where B is the saturation induction at room temperature. For a pure magnetic field: $U = 0.603 \times 10^{-7}$ eV/T.

As in optics, a neutron wave incident from vacuum on a plane surface of a medium will undergo total reflection at the boundary provided that the angle of incidence θ (as referred to the surface normal) lies in the range

$$\theta > \arcsin n. \tag{2.9}$$

This condition is equivalent to saying that the "perpendicular kinetic energy" $E_\perp = \hbar^2 k_\perp^2/2m$, corresponding to the normal component $k_\perp = k \cos \theta$ of the wavevector, should not exceed the scattering potential, i.e.

$$E_\perp < U, \text{ for total reflection.} \tag{2.10}$$

Since its first observation by FERMI and ZINN in 1946 /2.9/ total reflection of neutrons has been widely used, e.g. for the determination of scattering lengths, or for the purposes of polarization or neutron transport in "guide tubes" where neutron beams are confined in two dimensions /2.10, 11/. ZEL'DOVICH pointed out in 1959 /1.3/ that neutrons with energy E < U can be confined in three dimensions since they will be reflected nearly perfectly from suitable materials at any angle of incidence. It was demonstrated first in Munich /1.1/ and, with 10^3 times lower intensity, in Dubna /1.4/ that neutrons of the very low energies required for this type of confinement can be filtered out of thermal neutron spectra by the use of guide tubes. Confinement of UCN in closed material boxes was demonstrated first in Moscow by GROSHEV et al. in 1971 /2.12/. The possibility of using a magnetic barrier for neutron confinement, as proposed first by PAUL in 1951 /1.7/ and VLADIMIRSKY in 1961 /1.8/, has been demonstrated experimentally by KOSVINTSEV et al. in 1975 /2.13, 14/. An all-magnetic bottle, however, has so far not been tested.

2.2 Implications of the Low Energy of Very Cold Neutrons

2.2.1 Large Disorder Scattering

The energies of VCN are comparable with, or not much larger than, typical nuclear and magnetic scattering potentials U, in contrast to the situation for thermal neutrons. It follows from elementary scattering theory that VCN will thus be scattered very strongly by fluctuations of density, composition or magnetization in condensed matter. Specifically, the total elastic scattering cross section of inhomogeneities varies like 1/E over a wide range of energy. Furthermore, VCN are scattered into fairly large angles due to their long wavelengths. These scattering characteristics are the basis for the usefulness of VCN for the investigation of submicroscopic structures within and at the surface of solids and liquids /2.15 - 17/. Such applications will be discussed in greater detail in Section 4.2. They are, to a certain extent, complementary to neutron small-angle scattering (which has recently been reviewed by SCHMATZ et al. /2.18/).

2.2.2 Acceleration and Focusing by Gravity or Magnetic Fields

The interaction energy of a neutron with the earth's gravitational field and with magnetic fields is respectively 1×10^{-7} eV/m and 6×10^{-8} eV/T. Due to their low energy VCN can thus be effectively accelerated or slowed down utilizing gravity or magnetic fields. This is of interest for the production of UCN using vertical or inclined guide tubes. On the other hand, these effects offer the possibility to construct very simple and efficient spectrometers utilizing the special properties of the flight parabola or of magnetic six-pole fields for monochromatization and focusing. A magnetic six-pole spectrometer for the investigation of critical scattering has been proposed by GOLUB and CARTER /2.19/.

2.2.3 Spectral Transformation by Mechanical Devices

The velocity v of VCN lies within easy reach of mechanically attainable velocities. E.g., an energy of 10^{-4} eV corresponds to v = 138 m/s, and for E = 10^{-7} eV, v = 4.4 m/s. This is the basis for a method of efficient production of UCN by reflection of originally faster neutrons from moving mirrors. This principle of spectral transformation has been realized in the "neutron turbines" (or "mechanical UCN generators") which have been developed in Munich /2.20, 21/ and Sofia /2.22, 23/ (cf. Section 3.3).

2.2.4 The 1/v Law

In contrast to the properties of VCN discussed so far the large absorption cross section which follows the 1/v law is frequently undesired. It may be a serious problem in materials studies requiring relatively thick samples. On the other hand, it is of no significant consequence in thin-foil or surface investigations, nor in the proposed EDM and lifetime experiments. The large absorption is beneficial in precision measurements of nuclear capture cross sections /2.24 - 26/ (cf. Section 4.1.1).

2.2.5 Intensity Considerations

The fraction of VCN with energies $E < E_0$ in a Maxwellian spectrum at temperature T is $f = 1/2 \ (E_0/k_B T)^2$, where k_B is Boltzmann's constant. In all practical cases this fraction is quite small, e.g., for T = 300 K and $E_0 = 10^{-7}$ eV, $f \cong 10^{-11}$. An improvement by a factor of up to 10 was achieved by the use of suitable cold convertors /2.27/. The problem of low intensity is, however, not prohibitive, because VCN may be filtered very efficiently by the use of guide tubes, either directly or in combination with a spectral transformation. Moreover, special detectors with highly selective spectral sensitivity have been developed. As a result, very pure VCN beams could be obtained at an insignificant background and with sufficient intensity for all experiments of interest.

In fact, for many applications of VCN the experimental sensitivity is not determined by the absolute intensity, but by other factors like the density and useful
volumes in phase space, as in double-differential scattering experiments where for
given source and sample a theoretical gain factor ~1/E has been derived for the
scattered intensity /2.28/. In the proposed EDM and lifetime experiments the gain
due to the long neutron confinement time greatly outweighs the sacrifice of lower
intensity as compared to beam-type experiments.

Proposals have been made for high-intensity "super-thermal" UCN sources, based
on the preponderance of energy "down-scattering" over "up-scattering" in a particular system of polarized nuclei /2.29/ or in special converters at low temperature
/2.30/. Furthermore, the idea of UCN accumulation to high densities in pulsed reactors has been advanced /1.9/. The practical realization of all these suggestions,
however, would seem to be problematical.

3. Production, Guidance and Detection of Very Slow Neutrons

3.1 General Remarks

According to Liouville's theorem the phase-space density of an ensemble of systems
remains constant under any contact transformation. This theorem implies that it is
impossible to change the phase-space density of neutrons in a neutron beam by the
application of any exterior forces which act collectively on the set of particles
lying within any infinitesimal phase-space element under consideration (like gravity,
mirror reflection, macroscopic magnetic fields, etc). On the other hand, it is well
known that a gain in phase-space density, and hence in intensity, may be achieved by
the use of a cold moderator. (In this case the interaction is not collective, but
the scattering processes responsible for cooling the neutron spectrum act separately
on each single neutron). A cold source, however, requires a 4π illumination for full
efficiency, hence it is usually useful only when provided inside the reactor near
the core.

The phase-space density in a Maxwellian spectrum for a thermal equilibrium energy
E_T and a flux Φ,

$$\nu = \left[\Phi/(8\pi m E_T^2)\right]\exp(-E/E_T),\tag{3.1}$$

is practically constant over the energy range of VCN where the Boltzmann factor is
very nearly equal to 1. In consequence of Liouville's theorem the highest achievable
intensity in a VCN beam will thus be given only by the flux and temperature of the
primary thermal or cold neutron spectrum, regardless of spectral transformations by
gravity, etc.

However, this statement must be modified in the sense that if VCN are produced in thin converters, E_T is not related to the converter temperature T_C as under thermal equilibrium conditions (cf. Section 3.2). In fact, VCN are produced in a converter by a single inelastic scattering process, thus thermal equilibrium cannot be established. In "super-thermal sources" E_T may be even smaller than $k_B T_C$, and hence ν larger than the equilibrium value, at least in theory /2.30/. The transfer of neutron energy (and entropy) beyond the thermal equilibrium to another system may be accomplished also in other ways, e.g. using a nuclear-spin system as proposed by NAMIOT /2.29/.

3.2 Converters

Ultracold neutrons cannot be extracted directly from reflector or cold-source spectra by conventional beam tubes, because they fail to penetrate the structural wall material due to large absorption and, in many cases, total reflection. One solution of this problem consists in the use of originally faster neutrons, which do penetrate reasonably thin windows without significant losses, and their secondary deceleration to the UCN region by gravity or a "neutron turbine". Alternatively, a thin piece of moderator - a "converter" - may be provided inside the beam tube in order to replenish the low-energy tail of the Maxwellian spectrum below the cutoff due to the wall losses. Moreover, if the converter is cooled, it behaves similarly to a cold source with the result that the UCN yield is improved.

3.2.1 Theory of the Homogeneous Converter

The production of VCN by inelastic scattering of thermal or cold neutrons in a thin homogeneous slab of material has been considered in a number of papers /3.1, 1.9, 3.2 - 4, 2.27/. It is usually assumed that the slab thickness be small enough to ensure negligible attenuation of the incident spectrum but, on the other hand, large compared to the mean free path Σ_1^{-1} of VCN, where $\Sigma_1 = \Sigma_c + \Sigma_{heat}$ is composed of the macroscopic cross sections for nuclear capture and thermal heating. Then the differential flux of VCN may be written as

$$\phi(E) = s(E)/(4\pi\Sigma_1), \tag{3.2}$$

where

$$s(E) \cong \int_0^\infty \Sigma(E'{\to}E) \, \Psi \, (E')dE' = \Phi \, \overline{\Sigma_{cool}} \tag{3.3}$$

is the source density of VCN. $s(E)$ is determined by the total incident flux Φ and the cooling cross section $\Sigma(E'{\to}E)$, averaged over the incident spectrum $\Psi(E')$, which

is assumed to be isotropic. Eq.(3.2) shows that the VCN production is determined
by the total source strength $s(E)/\Sigma_1$ in an "effective surface layer" of the thickness (in the flight direction) of one mean free path of VCN, Σ_1^{-1}.

According to the 1/v law: $\Sigma_1 \sim E^{1/2}$, and from elementary phase-space considerations in inelastic scattering to very low energies: $\overline{\Sigma_{cool}} \sim E^{1/2}$. Hence $\phi(E) \sim E$.
This is the same behaviour as for the low-energy region of a Maxwellian spectrum.
It may be shown using the principle of detailed balance, that for an incident Maxwellian spectrum characterized by the temperature T_N, and a converter temperature
$T_C = T_N$, a converter without absorption would just reproduce the original spectrum,
i.e. re-establish its low-energy portion which was lost in the walls of the beam
tube. In the presence of absorption the intensity will be reduced to the fraction
$(1 + \Sigma_c/\Sigma_{heat})^{-1}$. Even for weak absorbers, this reduction is significant at low
temperatures, because for $T_C \to 0$, $\Sigma_{heat} \to 0$ (in consequence of the decreasing popu-
lation of excited states serving as a reservoir for up-scattering events). E.g.,
for a Debye spectrum at low temperature: $\Sigma_{heat} \sim T_C^{7/2}$, while for a two-level system
with level separation $\hbar\omega$, Σ_{heat} vanishes even exponentially, viz. $\Sigma_{heat} \sim T_C^{-2} \exp$
$(-\hbar\omega/k_B T_C)$.

These considerations allow a qualitative assessment of the variation of VCN yield
with T_C at fixed T_N: For $T_C \to 0$, $\Sigma_{heat} \to 0$, while $\overline{\Sigma_{cool}}$ remains finite, hence $\phi(E)$
should increase with decreasing T_C until absorption begins to dominate. The crite-
rion for a good choice of T_N is that there should be a broad overlap of the incident
neutron energies with the spectrum of converter excitations in order that the VCN
energy region may be reached in a single down-scattering event.

It should be emphasized that the VCN spectrum for $T_C \neq T_N$ is not in equilibrium
with the converter temperature in general. The calculations of PORSEV and SEREBROV
/3.4/ show that for T_N = 400 K, the VCN flux from Be should even significantly exceed the equilibrium flux in the range T_C < 200 K (and also for T_C > 400 K). GOLUB
and PENDLEBURY /2.30/ have pointed out that especially high gains above thermal
equilibrium could be achieved in a converter system with a two-level excitation
spectrum at low temperature.

The variation of theoretical VCN flux with T_C and T_N has been calculated for a
number of materials of practical interest, usually in the form of a gain factor
$G(T_N,T_C)$ referred to polyethylene at 300 K, which practically corresponds to a Maxwellian at 300 K. GOLUB considered hydrogenous media assuming a Debye phonon spectrum
/3.2/. GOLIKOV et al. /3.3/ and PORSEV and SEREBROV /3.4/ calculated $G(T_N,T_C)$ for
solids like polyethylene, beryllium, aluminium, magnesium, and zirconium hydride,
using experimental phonon spectra. AKHMETOV et al. /2.27/ analysed various gaseous
and frozen converters (H_2O, D_2O, H_2, D_2, $C_{15}H_{16}$). The calculated gain is particularly high for para-hydrogen where $G \cong 30$ for T_C = 20 K and T_N varying over a wide range
from 100 to 300 K.

3.2.2 Real Converters

Many hydrogenous substances, like methane, cannot be used as converters, in spite of their favourable nuclear and dynamic properties, because of insufficient chemical stability in intense radiation fields. Polyethylene has been used successfully both in the low-power IBR-30 reactor in Dubna /3.3/ and in a fairly high thermal flux of 10^{12} cm^{-2}s^{-1} in the thermal column of the NRU reactor at Chalk River /3.5, 6/ where the fast flux is very low. Polystyrene is in use in a thermal flux of 10^{12} cm^{-2} s^{-1} at the Universities Reactor, Risley /3.7/. The only hydrogenous material which has so far been applied successfully in a high-flux reactor, however, is zirconium hydride which was found to withstand, without casing, a thermal flux of $(2-4) \times 10^{14}$ cm^{-2}s^{-1} at the SM-2 reactor, Dimitrovgrad /3.8/.

A number of good nonhydrogenous converter materials, like Be, graphite, or D_2O, are not suitable for horizontal UCN guide tubes due to their large scattering potential U, since the value of U constitutes the low-energy cutoff for the emergent VCN spectrum. In a vertical or inclined beam tube, on the other hand, refractive effects are irrelevant owing to the secondary gravitational neutron deceleration (see Section 3.4). Hence graphite and Be could be used successfully in vertical UCN facilities in Munich /1.1, 3.1/ and Leningrad /3.9/, respectively. Accordingly, vertical or inclined arrangements have also the advantage that window losses due to casings of gases, liquids, or volatile solids (like Mg), become insignificant. Even a simple vertical arrangement with a thin window at the beam tube nose provided instead of a converter gave very satisfactory results /3.10/.

In addition to possible window and refraction losses there exists, in disordered solids used as converters, another effect which may cause a significant reduction of the VCN current $I(E,\theta)$ from the converter below the cosine distribution $I_0(E,\theta)$ = $\phi(E)\cos\theta$ which holds for homogeneous substances. (The polar angle θ refers to the surface normal). This reduction is due to the strong scattering of VCN from the fluctuations of the scattering-length density associated with any inhomogeneities comparable in size to the neutron wavelength. Such elastic scattering gives rise to a decrease of the VCN mean free path, and hence of the effective source volume for the emergent neutrons. This effect has been analyzed in /3.1, 2, 10/, assuming an isotropic distribution of elastic scattering. The result of /Ref. 3.10, footnote 26/ may be written in the form

$$I(E,\theta)/I_0(E,\theta) = H(\cos\theta)\ \sqrt{1-\alpha}, \tag{3.4}$$

or, for the angle-integrated current $J_{(0)}(E) = 2\pi\int_0^1 I_{(0)}(E,\theta)\ d(\cos\theta)$,

$$J(E)/J_0(E) = 2\ \phi_1\ \sqrt{1-\alpha} \tag{3.5}$$

α is the ratio of the macroscopic elastic scattering cross section Σ_s to the total cross section $\Sigma_1 + \Sigma_s$. $H(\cos \theta)$ denotes the tabulated H-function, which depends on α and θ, and Φ_1 is its first moment, $\Phi_1 = \int_0^1 H(\mu)\, d\mu$ (see, e.g. /3.11, 12/). In the strongly inhomogeneous substances electrographite and hot-pressed Be, e.g., the calculated reduction of the total UCN current amounts to a factor of ≈ 4 for graphi at 500 K, and of ≈ 6 for Be at 80 K.

Systematic experimental studies of various converter materials at different temperatures have been performed in Dubna /3.3/, Moscow /3.13/, Alma Ata /3,14, 2.27/, Chalk River /3.6/, and Grenoble /3.15/. The substances investigated in Dubna, Mosco and Chalk River were polyethylene, Al, zirconium hydride, Mg (canned in an Al box), and light water. Mainly gaseous and frozen converters were tested in Alma Ata. As a example, Fig.3.1 shows the scheme of the gaseous converters tested in a horizontal through tube at the VVR-K reactor at Alma Ata /2.27/ in a thermal flux of 5×10^{12} $\mathrm{cm}^{-2}\mathrm{s}^{-1}$. The authors used H_2, D_2, He and air. Cooling was achieved by circulating liquid N_2 or tap water. The copper foil lining serves as a reflector for UCN. It wa provided as a means to effectively double the converter volume. The experimental results show, e.g., a gain factor of 3.2 for H_2 on cooling from 300 K to 80 K, in agreement with the calculation for a thermal equilibrium concentration ratio of ortho-H_2 to para-H_2.

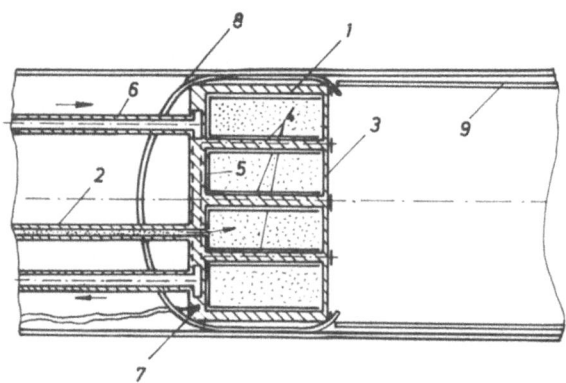

Fig.3.1. Scheme of the gaseous converter for ultracold neutrons at the reactor VVR-I Alma Ata /2.27/. 1 aluminium box; 2 gas inlet tube; 3 window (Al 0.25 mm) for neutr(exit to the guide tube; 4 supports; 5 Cu foil; 6 tubes for the cooling system; 7 thermocouple; 8 spacer spring; 9 guide-tube

In the same set of experiments the authors also investigated thin frozen H_2O, D_2O and alcohol films deposited from the vapour phase on a cold aluminium plate. Expecially favourable results were reported for H_2O films at 80 K where the intensity was found to be higher by a factor of ≈ 4 than with H_2 at 80 K. The gain factor as compared to H_2O at 300 K was determined to be 3.4, which is ≈ 60 % of the calculated gain due to cooling. Similar gains and deviations from expectation had been observed previously by GOLIKOV et al. /3.3/ also for polyethylene and zirconium hydride. The authors suggest that inadequacies at low energies of the phonon spectra used in the calculations and/or converter inhomogeneities may possibly account for the discrepancies.

The experiments at the Institut Laue-Langevin, Grenoble /3.15/ indicate that a very large gain factor of. 30, in agreement with theory, can be obtained with para-hydrogen at 18 K, while all investigated solids (H_2O ice, Be, polyethylene, graphite) show strongly reduced gains, apparently due to inhomogeneities.

Due to the large uncertainties of determinations of transmission losses in guide tubes for very low energy neutrons, and of detection efficiencies, it is usually quite difficult to derive reliable absolute values for the converter efficiency from the measured intensity. Therefore, only the characteristic parameters and overall performance of various UCN installations will be given later in Table 3.1, Section 3.6.

3.3 Mechanical Generators of Ultracold Neutrons

Prior to the construction of the first UCN facility, the vertical spectrometer for very low energy neutrons at Munich /1.1/, an alternative method of producing UCN by a "neutron turbine" had been pointed out by STEYERL in 1966 /2.20, 3.16/. The "neutron turbine" is based on the possibility of using a spectral transformation of slow neutrons to reach the energy range of VCN. TUNKELO and PALMGREN had demonstrated in 1967 /3.17, 18/ that such a transformation is feasible, using a sample moving away from the incident neutrons at a speed close to the neutron velocity. In contrast to Tunkelo and Palmgren's design, the turbine offers a way to achieve a neutron deceleration in the laboratory system, and in this way to provide a continuous beam of UCN.

Such a turbine has been constructed and successfully tested at the FRM reactor at Munich /2.21/. Its principle of action is represented in Fig.3.2. Neutrons with an original verlocity v_1 ≅ 50 m/s, corresponding to a wavelength of 80 Å, enter the system of curved neutron mirrors, made of densely spaced thin copper shells, moving at a speed v_T = 25 m/s. The neutron direction of flight is nearly reversed in the moving system by about 10 reflections along the well-polished "blades" In this way a deceleration by twice the blade velocity to a final velocity near zero

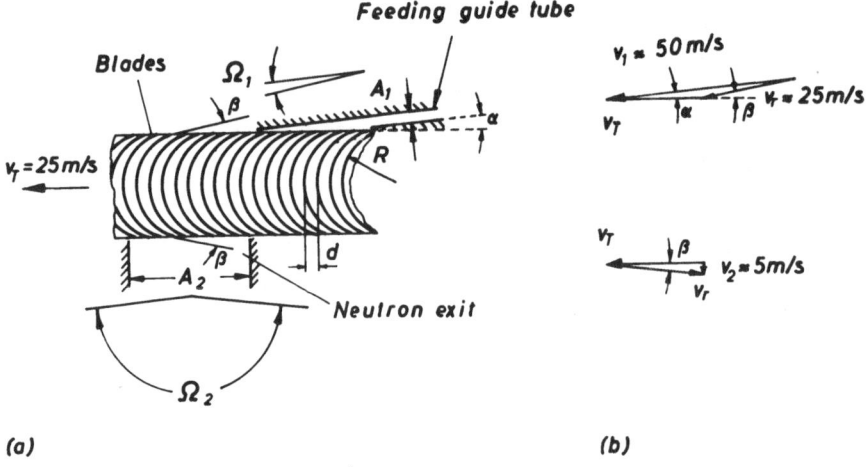

(a)   (b)

<u>Fig.3.2 a and b.</u> Principle of the "neutron turbine" constructed at the FRM reactor, Munich /2.21/. Neutrons provided by a guide tube are decelerated by several total reflections from the moving curved blades which are arranged along the circumference of a rotating wheel (a). In this process the beam cross section (A_1 and A_2) and the divergence (Ω_1 and Ω_2) increase. The velocity triangles at the entrance and exit (b) illustrate the deceleration from an original velocity v_1 to a final velocity v_2 (v_T: blade velocity; v_r: velocity relative to the blades)

in the laboratory system is accomplished ($v_2 \lesssim 10$ m/s). In accordance with Liouville's theorem the primary beam of slow neutrons experiences, during deceleration, a considerable broadening both in cross section (by a factor of 7 to a final cross section of $\approx 10 \times 20$ cm^2) and in divergence (to an isotropic distribution over the full solid angle 2π). Such a large, diffuse source of UCN is desired for most applications of very low energy neutrons.

The most important practical advantage of a mechanical generator of UCN over stationary UCN channels seems to be the possibility to avoid large installations near the reactor core and, instead, to use a simple, narrow guide tube, which does not necessarily require a converter, for feeding the turbine with cold neutrons. The turbine may be installed close to the experiments; thus the considerable losses occurring in UCN channels (see Section 3.4) may be eliminated. The total turbine loss due to imperfect mirror reflections from the blades was measured to amount only to a factor of two.

Similar proposals for cold-neutron decelerators have also been advanced by ANTONOV et al. /3.19/ and by KASHUKEEV /2.22, 3.20/. Kashukeev's device has been constructed and tested at the Institute of Nuclear Research and Nuclear Energy, Sofia

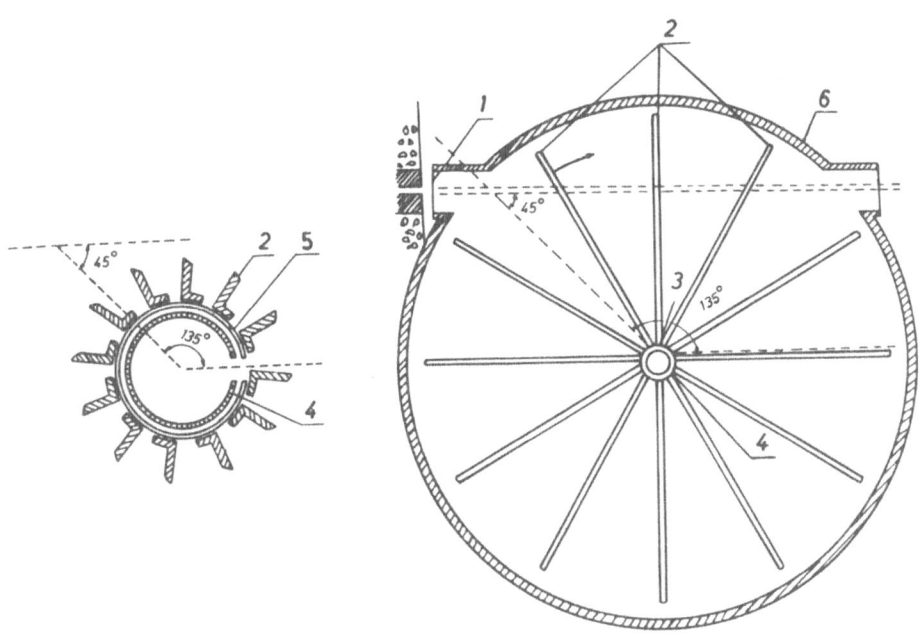

Fig.3.3. Scheme of the mechanical generator of ultracold neutrons constructed at
the Institute of Nuclear Research and Nuclear Energy, Sofia /2.23/. Very slow neutrons enter the rotating system through window *1*. Those satisfying the condition
for total reflection on their first collision with a paddle *2* bounce down the paddle
to the hub *3*. In this process they are decelerated to the region of ultracold neutrons by the action of the centrifugal force and enter a neutron channel *4* placed
inside the hub. *5* Cd shield; *6* steel casing

/3.21, 2.23/. Fig.3.3 illustrates the scheme of Kashukeev's design which is based
on the principle of the "radial turbine", in contrast to the "axial turbine" at the
FRM as described above. An advantage of the radial solution is the possibility of
stronger deceleration, because the centripetal potential is utilized. Disadvantages
are higher reflection losses due to a greater number of reflections along the pad-
dles, and incomplete angular dispersion in the sense that the beam entering the UCN
channel is not isotropic but collimated. Such a beam cannot fill the channel isotropically up to the theoretical density corresponding to the phase-space density
in the beam, since the effective aperture for neutrons leaving the channel through
the entrance hole is larger than for the incoming neutrons.

The above techniques of neutron deceleration by reflection from moving mirrors
utilize total external reflection. On the other hand, Bragg reflection from rotating

or reciprocating monocrystals has been widely used with thermal neutrons to obtain pulsed monochromatic and/or Doppler-shifted neutron beams /3.22/. BALLY et al. have built a moving-crystal monochromator /3.23/ at the Institute for Atomic Physis, Bucharest, which is expected to permit a Doppler shift down to the velocity region of ▪ 100 m/s /3.24/. Because the process of Bragg reflection is highly selective in momentum space, such a device would, however, seem to be useful only for applications which require a very high resolution.

3.4 Propagation of Very Low Energy Neutrons

Since the pioneering work of MAIER-LEIBNITZ and SPRINGER /2.10/ it is well known that slow neutrons may be transported over sizable distances at low losses, by the use of "neutron guide tubes" along which the neutrons are conducted by multiple total reflection from the walls. Extensive use of guide tubes has been made in all existing facilities for very low energy neutrons.

In the analysis of guide transmission it is important to note, that the mechanism of propagation is somewhat different for ultracold neutrons as compared to very cold neutrons, in consequence of the peculiar property of UCN to experience total external reflection at any angle of incidence on the channel walls, whereas neutrons of higher energy are reflected only if incident within a limited range of glancing angles (see Section 2.1). Therefore, UCN may propagate along the channel similarly as the molecules in a rarefied gas, while the conduction of faster neutrons resembles that of light in a light pipe.

Accordingly, the mechanism of beam attentuation along the guide tube is quite different for UCN and for neutrons of higher energy. UCN are thought to disappear from the guide tube practically only as a result of absorption or inelastic scattering in the reflecting surface layer of the wall, or through the entrance or exit opening. The predominant loss mechanism for faster neutrons, on the other hand, is diffuse, nonspecular reflection due to surface roughness, with the result that the neutron leaves the angular range suitable for further total reflection.

The diffuse reflection of slow neutrons from slightly uneven surfaces has been analysed by STEYERL /3.25/ both for submicroscopic roughness and for macroscopic surface waviness. In either case the beam suffers an angular spreading about the direction of mirror reflection for a plane surface. It was shown that in the region of very low neutron energies the effect of submicroscopic surface roughness contributes overwhelmingly to the losses in guide tubes for any technically realistic surface qualities. For example, the average loss per reflection for VCN with wavelength 100 - 400 $\overset{\circ}{A}$ was calculated to be $\approx$ 4 % for a mechanically and electrolytically polished copper tube for which the parameters of microroughness were reported in /3.1/ to be: a $\cong$ 35 $\overset{\circ}{A}$ for the RMS amplitude of asperities, and w $\cong$ 250 $\overset{\circ}{A}$ for the lateral correlation length. Such a value for the reflection coefficient is in good agreement with the experimental value of ≈ 3 % for the vertical VCN spectrometer in Munich /3.1/. A somewhat smaller loss of ≈ 2 % per reflection was measured for honed and electropolished stainless steel tubes /3.26, 27/. For good mirror glass, where a ≅ 10 Å, a value below 1 % is expected.

The transmission losses of UCN through guide tubes seem to be not as well understood. The theoretical absorption probabilities for the UCN wave in the reflection process are so small for most substances (e.g., 2.5×10^{-4} for Cu) that it was originally thought UCN would diffuse almost ideally along any tube, irrespective of its geometry and surface quality. However, the later experiments on UCN containment in traps yielded much higher reflection losses than expected (see Section 5). Accordingly, the attenuation of UCN intensity along horizontal channels was observed to be quite significant in all existing facilities (except for the first one of this type in Dubna /1.4/ where the data taken at an extremely low intensity had apparently been interpreted in an overoptimistic way). A considerable improvement of UCN transmission was achieved by avoiding sharp channel bends, which were shown to be highly detrimental by measurements of KOSVINTSEV et al. /3.8, 28/, and by a better surface quality, using electropolished Cu or stainless steel tubes in order to enhance specular reflection. By such means the mean number of reflections which a neutron needs to travel through a tube of given length, could be reduced very efficiently.

An analysis of UCN propagation has been reported first by LUSHCHIKOV et al. /1.4/. The authors used elementary diffusion theory, assuming completely diffuse scattering of UCN in collisions with the walls. Later the diffusion approach was modified by GROSHEV et al. /2.12/ and SHAPIRO /1.9/ in order to take account of the partial specularity of reflections from polished walls, leaving the diffusion length and the diffusion coefficient as quantities to be determined experimentally (from the variation of intensity with the channel length and with the area of an absorber covering a part of the channel). From the experimental results of GROSHEV et al. /2.12/ for an electropolished copper tube with internal diameter 9.4 cm the authors determined the diffusion length, which describes the distance for a (1/e) drop of intensity, to be 4.7 m. It was further concluded /1.9/ that the probability of specular reflection was 82 %, assuming that the nonspecular fraction of reflected intensity follows the cosine law.

The same scattering distribution was also used in Monte Carlo calculations of UCN flow by WINFIELD and ROBSON in 1975 /3.29/. ROBSON reported 90 % specularity for a mechanically polished copper tube /3.6/, on the basis of these calculations. However, it had been shown previously by STEYERL in 1972 /3.25/ and IGNATOVICH in 1973 /3.30/ that for UCN scattered from microroughness such a perfectly diffuse scattering distribution is a poor approximation. BERCEANU and IGNATOVICH presented in 1973

/3.31/ a modified analysis of UCN diffusion through a guide tube, using as an improved approximation for the distribution of nonspecular reflection the form
$\cos\theta_0\cos^2\theta$, where θ_0 and θ respectively denote the polar angle of incidence and
reflection. They also showed that their analysis of transmission probability and
angular distribution of transmitted neutrons was in agreement with Monte Carlo simu-
lations, at least for the considered case of zero absorption. BROWN et al. reported
in 1975 /3.32/ on Monte Carlo calculations using the exact theoretical expression
for the scattering distribution ($\sim\cos\theta_0\cos^2\theta \, \exp(-\kappa_{\shortparallel}^2 w^2/2)$, where w is the lateral
correlation length for the asperities and $\hbar\kappa_{\shortparallel}$ the component of momentum transfer par
allel to the wall surface). Further insight into the effect of roughness was gained
by the experimental study of the nonstationary diffusion of a pulsed UCN beam througI
a section of guide tube by EGOROV et al. /3.9/ and by the theoretical analysis of
such a process in the framework of diffusion theory by VINOGRADOV and TEREKHOV
/3.33/. A nonstationary process was also studied experimentally and theoretically
by ROBSON /3.6/. In spite of differences in details, all these investigations reveal
that nonspecular reflection not only leads to beam peaking in the axial direction,
but also must be considered to be the major source of attenuation of intensity along
guide tubes for UCN.

Thus, the basic requirements for a good UCN channel have eventually turned out to
be quite the same as for the guide tube for faster neutrons: high surface quality
ana use of bends only to the extent necessary for suppressing the background due to
γ-rays and undesired neutrons of higher energy. In some cases sharp bends were pre-
ferred to smooth bends, in order to provide a very pure beam of UCN at the expense
of increased losses.

3.4.1 Vertical Guide Tubes

We describe first the spectrometer for very low energy neutrons built in Munich in
1967/68 on the proposal of Maier-Leibnitz, where in early 1968 neutrons with energies significantly below 10^{-4} eV and reaching down to the UCN region have, for the
fjrst time, been observed and used to study their interaction with matter /1.1/.
This facility is also unique in respect to the wide energy range provided ($10^{-7} \leq$
$E \leq 5 \times 10^{-4}$ eV) and to its special chopper system which allows time-of-flight (TOF)
spectroscopy in a rather efficient way.

Fig.3.4 shows the scheme of this facility /1.1, 3.1, 10/. It consists of a slightly S-shaped, highly polished neutron guide tube (internal diam. 5 cm) and a rotating
Fermi-type chopper made of parallel thin glass plates. Since the angular range of
transmission through such a chopper increases proportionally to wavelength λ, low-
energy neutrons are favoured and the TOF resolution becomes nearly independent of
λ ($\approx$ 5 - 20 %, depending on the chopper frequency).

The intensity measured at this spectrometer was found to be in agreement with a
Maxwellian distribution if allowance is made for the losses in the guide tube and

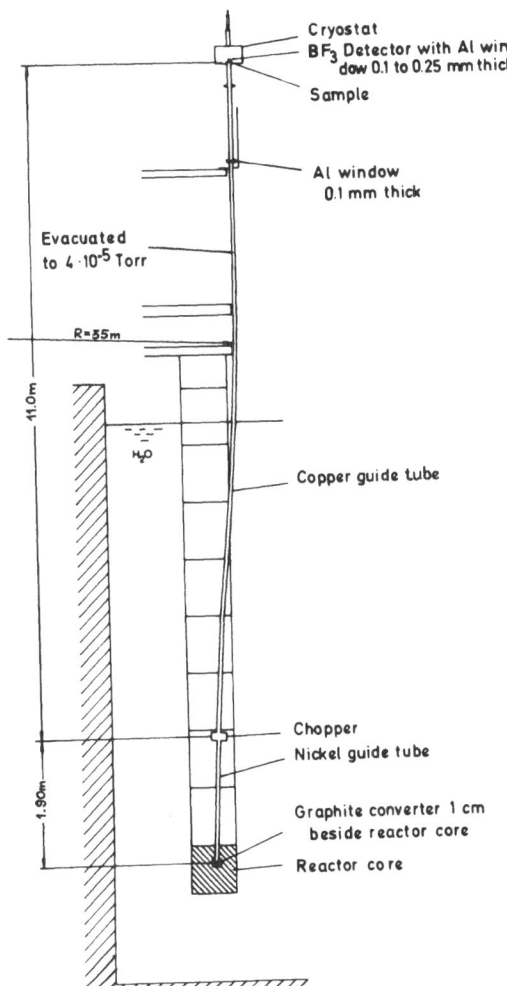

Fig.3.4. Scheme of the vertical time-of-flight spectrometer for very low energy neutrons constructed at the reactor FRM, Munich /1.1, 3.1/. The nickel-coated aluminium tube was later replaced by an electropolished copper tube. In addition, the graphite converter was removed and instead a thin aluminium window installed /3.10/

the chopper. These losses increase strongly with λ. For example, the total reflection losses in the guide tube amount to a factor of 2.5 at 80 Å and to a factor of 8 at 1000 Å. Due to the very efficient suppression of background by the curvature of the guide tube, however, measurements could be performed down to neutron velocities of $v \cong 4$ m/s ($\lambda \cong 1000$ Å). For example, in the interval $4.5 < v < 5.5$ m/s a UCN beam intensity of ≈ 1.4 neutrons/cm^{2}s was observed. Although such a neutron current corresponds to only ≈ 3 % of the primary Maxwellian distribution, it still exceeds the background by ≈ 100 times.

Another vertical guide tube designed for UCN has been constructed at the VVR-M reactor of the Institute of Nuclear Physics, Leningrad, by a group working under Professor Lobashov /3.9/. Fig.3.5 shows the arrangement of this installation. The counting rate of UCN (as determined by transmission through the nickel screen 5) obtained with an uncooled Be converter ($T_c \cong 400$ K) in a thermal flux of $(3 \times 5) \times 10^{13}$ cm^{-2}s^{-1} was 1500 s^{-1} for the total channel cross section of 7×6 cm^2, i.e. 36 neutrons/cm^{2}s (see Table 3.1). This value indicates that the overall loss factor for UCN extraction is only about 10 for this beam tube, i.e. 2.5 times lower than for the vertical guide tube at Munich (where the beam obstruction due to the chopper is very significant at the lowest energies).

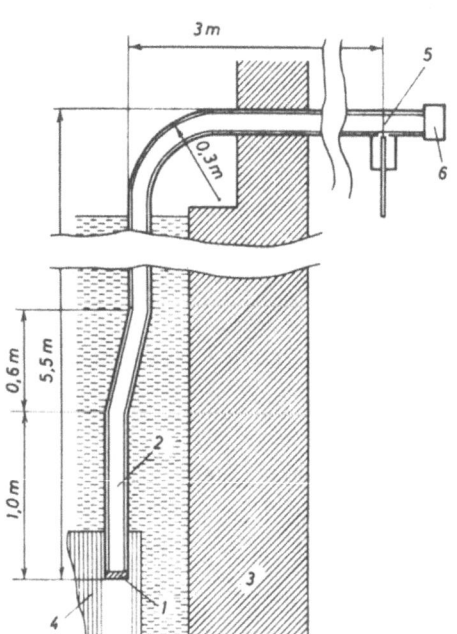

Fig.3.5. Arrangement of the vertical channel for ultracold neutron extraction constructed at the reactor VVR-M Leningrad /3.9/.
1 beryllium converter; 2 mirror channel; 3 reactor shielding; 4 reactor active zone; 5 Ni screen for determining the background (of neutrons with velocities exceeding 6.7 m/s); 6 detector.

The portion of the neutron guide up to the smooth 90° bend is made of stainless steel. The smooth bend and the horizontal part consist of glass plates with deposited ^{58}Ni

An inclined beam tube for very low energy neutrons ($E \gtrsim 10^{-4}$ eV), containing an H_2O converter and guide tubes made of stainless steel and nickel-coated glass, is presently under construction at the high-flux reactor of the Institute Laue-Langevin, Grenoble.

3.4.2 Horizontal Channels for Ultracold Neutrons

The first horizontal channel for UCN has been built in 1968, upon the initiative of Professor Shapiro, at the pulsed IBR reactor of the Joint Institute for Nuclear

Research, Dubna, where the mean thermal flux at the converter was only $\simeq 10^{10}$ cm^{-2}s^{-1} /1.4/. Accordingly, the detected UCN flux was extremely low (2 x 10^{-4} cm^{-2}s^{-1}). Significant intensities of up to 1.6 cm^{-2}s^{-1} were reported later by GROSHEV et al. /2.12/ and SHAPIRO /1.9/ for similar installations at the IRT-M reactor of the Institute of Atomic Energy, Moscow. The gain was due mainly to a much higher thermal flux (up to 2 x 10^{13} cm^{-2}s^{-1}) and to improved surface quality of the Cu guide tubes achieved by electropolishing. In addition, better converter materials like H_2O, Mg, or zirconium hydride, were used instead of aluminium, and in some experiments, the early scintillation counters were replaced by special ^{3}He detectors with higher efficiency of UCN (see Section 3.5).

A remarkable improvement over all-copper guide tubes was achieved more recently in other UCN facilities by using, as a guide tube material near the core, stainless steel which is more stable against radiative corrosion than copper, as the experiments of LOBASHOV et al. /3.9/ have shown. In addition, stainless steel is capable of retaining a larger section in momentum space, since its critical velocity for total reflection, v_{cr} = 6.2 m/s, exceeds that for copper (v_{cr} = 5.6 m/s). The mirror reflectivity of well polished stainless steel is also superior to that of Cu /3.26,, 27/. Furthermore glass has been used for the out-of-pile part of UCN guides /3.7/.

A number of horizontal UCN channels featuring such improvements have been built in recent years: At the Universities Reactor, Risley /3.7/, at the NRU reactor, Chalk River /3.5, 6/, at the VVR-M reactor, Leningrad /3.9/, at the VVR-K reactor, Alma Ata /2.27, 3.14/, and at the high-flux reactor SM-2, Dimitrovgrad /3.34, 8/. As an example, Fig.3.6 shows the scheme of the installation at the SM-2 reactor, where the thermal flux at the position of the converter (zirconium hydride) is as high as (2 - 4) x 10^{14} cm^{-2}s^{-1}. A total counting rate of 640 s^{-1} (corresponding to a flux of 11 neutrons/cm^{2}s) was reported for UCN below the copper threshold of 5.6 m/s. The differential spectrum of UCN was measured by means of a gravitational spectrometer which allows to change the height of neutron rise in a rotary guide tube elbow placed before the detector, as described first by GROSHEV et al. /2.12/. Such measurements confirm that the UCN spectrum corresponds to the initial region of a Maxwellian spectrum.

3.5 Detectors for Very Low Energy Neutrons

Detectors for very cold, and especially for ultracold, neutrons should meet a number of special requirements related to the particular features of very low energy neutrons, like large absorption cross section, refraction and low intensity:

a) Entrance windows, where necessary, should be made of thin, weakly absorbing material with a low scattering potential.

b) The quantity of reactive material used for neutron detection by neutron-nucleus reactions (e.g., ^{10}B(n,α), ^{3}He(n,p), ^{6}Li(n,α), or ^{235}U (fission)), may be kept much

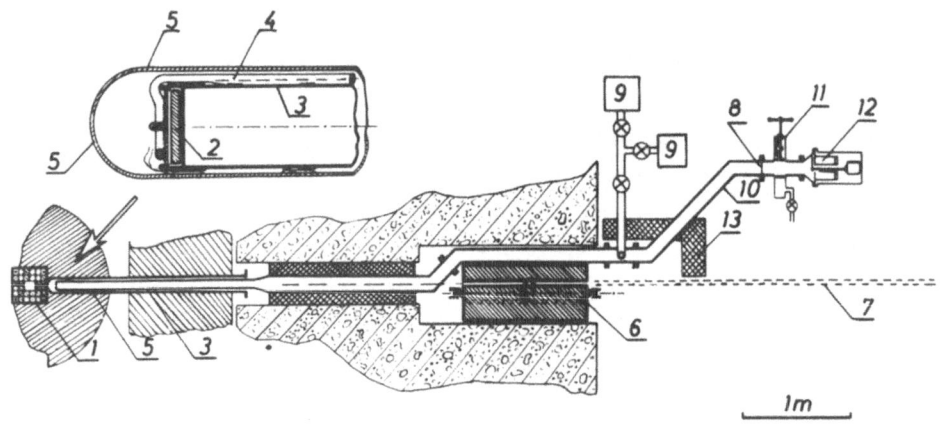

Fig.3.6. Scheme of the horizontal ultracold neutron channel at the reactor SM-2, Dimitrovgrad /3.8/. 1 reactor active zone; 2 converter (zirconium hydride); 3 neutron guide (electropolished stainless steel); 4 water cooling system for the converter; 5 vacuum jacket (zirconium); 6 shutter for direct beam; 7 direct beam; 8 aluminium window; 9 oil-free vacuum pumps; 10 neutron guide (electropolished copper); 11 vacuum valve; 12 ^{3}He detector; 13 additional biological shielding

smaller than for thermal neutron detectors. This offers the possibility to suppress the background from faster neutrons very efficiently by carefully limiting the absorber thickness.

c) A limit on the choice of absorber substance is imposed by the requirement that the very slow neutrons should be able to penetrate it. Therefore, only compounds with low scattering potential may be used for UCN detectors.

d) A further special feature of many UCN detectors is their large sensitive area as required for full utilization of the large cross sections of typical UCN sources.

e) The experimental conditions of some UCN containment experiments, where the neutrons were collected near the reactor active zone, demanded the development of special detectors with a particularly high stability and discrimination in intense radiation fields.

According to such requirements a large variety of detector systems have been developed by the various groups working with very low energy neutrons: Gas-filled proportional counters (BF_3, ^{3}He), scintillation detectors, fission counters, activation foils, track analyzing systems, radiographic techniques, and others.

Among these, proportional counters have seemingly turned out to be superior in many applications with respect to such criteria as high efficiency, reliability, low background, and simplicity. BF_3 counters with thin Al or Ti windows have been employed in Munich /3.1/ and Chalk River /3.6/. The overall efficiency of a BF_3 detector with reduced ^{10}B concentration (11 %) and a 0.1 mm thick Al window for a divergent beam of UCN below 5.6 m/s was estimated to be about 50 % /3.10/. About the same efficiency was reported for a special ^3He detector which was developed in Dubna by STRELKOV /3.35, 13/. The special features of this detector are a very low ^3He concentration (0.6 % in 97 % Ar and 2.4 % CH_4 or CO_2), totally reflecting internal surfaces (except for the thin Al entrance window), a large area (2 x 30 cm^2) and optimal uniformity of response across it. An improvement of window transmission by rotation of a corrugated window or by using cold polyethylene which has a scattering potential near zero (while for Al, U = 0.054 μeV) is apparently possible /3.36/.

Windows may be avoided entirely in solid scintillation detectors as developed first in Dubna /1.4/. The active substance of this type of counter is usually a thin film of ^6Li containing compound, e.g. $LiOH \cdot H_2O$ on ZnS. Improved versions were reported by LOBASHOV et al. /3.9/ and by BATES and ROY /3.37/. For the purpose of better discrimination against light particles, ANTONOV et al. used the fission reaction of ^{235}U, thermo-diffused into Ti in order to reduce the albedo of ^{235}U /3.38/. These authors tried various scintillators and achieved very low sensitivity to background radiation by using ZnS(Ag) or gaseous xenon /3.39, 40/. Generally, the efficiency of scintillation counters was found to be lower than, or at most comparable to, that of BF_3 or ^3He detectors.

Since all these detector systems act in a prompt way they can be used for non-stationary processes. Some other detection systems lack this property: ROBSON and WINFIELD used the activation of a thin manganese foil for UCN detection /3.5/; ANTONOV et al. developed a highly selective and efficient track detector where the fission products of ^{235}U are intercepted by a mica plate /3.41/, and BATES and ROY reported on radiography with very cold neutrons /3.7/.

Finally, a convenient but less efficient way of UCN detection should be mentioned, where the detection is preceded by inelastic scattering to thermal energies /3.10, 36/.

3.6 Survey over Various Facilities for Very Low Energy Neutrons

Table 3.1 comprises characteristic data for a number of sources for very low energy neutrons. In some cases the best performance achieved at a specific installation under various conditions, regarding the converter, guide tube, detector etc., was listed.

The figure of merit given was calculated on the basis of a comparison of the phase-space density of detected neutrons with the phase-space density corresponding to the low-energy region of a Maxwellian spectrum of primary thermal neutrons. Thus, for a thermal converter or in the absence of a converter the figure of merit is a direct measure of the overall effect of the transmission loss in the guide tube, of the detector efficiency, and possibly other factors like the turbine conversion efficiency. For a cooled converter where the phase-space density of the source is increased, the figure of merit will be higher, and would exceed 1 in the absence of any losses.

It is estimated that the figure of merit is subject to an uncertainty of at least a factor of two, due to uncertainties in the primary flux, the detection geometry, and other factors. Furthermore, since the figure of merit is referred to the thermal flux bathing the converter, the flux depression caused by the UCN installation itself is left out of consideration. The flux depression is very significant for some of the large installations near the reactor core. In spite of these shortcomings the figure of merit allows a semiquantitative judgement of the various methods of UCN production. Table 3.1 shows that some UCN sources are definitely more efficient than others, but that the total losses are fairly high for all existing techniques of UCN production.

4. Interaction of Very Cold Neutrons with Matter

A unique feature of slow neutrons, as compared to any other elementary particles, is their availability at very low energies. This allows one to study the neutron interaction with matter at extremely low impact energies. It is well known that, on the basis of very general considerations, quantum theory makes the following specific predictions on the scattering and reaction behaviour of matter waves in the limit of long wavelengths (see, e.g. /4.1/):

a) The total cross section for elastic scattering reaches a constant value given by a "scattering length" if the interaction potential as a function of distance is of shorter range than $1/r^3$.

b) All reaction processes, where the internal state of any of the colliding systems is changed, follow the $1/v$ law provided that the interaction potential decreases more strongly than $1/r^2$. It should be emphasized that we are obviously allowed to include in this definition of "reaction processes" not only any particle-induced nuclear reactions, but also the collective inelastic scattering from a system of atoms as described, e.g., by phonon scattering. Even in this case the $1/v$ law should be valid as long as the particle-nucleus interaction is of sufficiently short range.

Very slow neutrons have been used to put these predictions to test in an energy region which is practically inaccessible to any other scattering particles.

Furthermore, since the wavelength of very cold neutrons is of the order of 10^1 to several times 10^2 Å, such neutrons are of interest to study structures with sizes

Table 3.1. Characteristics of various sources for very low energy neutrons

Installation Institute	Converter, Temperature, Thermal Flux (neutr./cm²s)	Neutron Guide Tube (d = diam., ℓ = length)	Neutron spectrum	Detector and its area	Detected Intensity (neutr./cm²s) for the given spectral range	Figure of merit	References
Vertical TOF spectrometer FRM, Munich	No converter, 0.5 mm Al window (previously graphite 500 K) 1×10^{13}	Cu, d = 5 cm, ℓ = 13 m (previously Ni-coated Al near the core)	8×10^{-8} < E(eV) < 5×10^{-4}, 10 % TOF resolution	BF_3 (0.1 - 0.25 mm Al window) 7.5 cm²	e.g., 0.8 in the range v = 5.7 ± 0.5 m/s	0.04	/3.1, 10/[a]
Horiz. Channel, Inst. of Atomic Energy, Moscow	H_2O in Al shell (0.3 mm), 300 K, 2×10^{13}	Cu, d = 9.6 cm, ℓ = 6 m	v < 5.6 m/s	^3He (0.06 mm Al window) 60 cm²	1.7 (v < 5.6 m/s)	0.02	/1.9/[a]
Horiz. Channel, NRU, Chalk River	Polythene 330 K 2×10^{12}	Cu, d = 7.3 cm, ℓ = 5.3 m	v < 5.6 m/s	Activation of Mn foil, 40 cm²	0.7 (v < 5.6 m/s)	0.08	/3.5, 6/

Table 3.1 (continued)

Installation Institute	Converter, Temperature, Thermal Flux $(neutr./cm^2 s)$	Neutron Guide Tube (d = diam., ℓ = length)	Neutron spectrum	Detector and its area	Detected Intensity $(neutr./cm^2 s)$ for the given spectral range	Figure of merit	References
Horiz. Channel, Univ.Reactor, Risley	Polystyrene ≈ 300 K 2×10^{12}	Glass 15×5 cm² $\ell = 8$ m	$2.5 < v < 19$ m/s	^{6}LiF on ZnS, 62 cm²	0.3^b $(v < 4.3$ m/s$)$	0.10	/3.42/
Horiz. Channel, Inst. of Atomic Phys., Alma Ata	$ZrH_{1.9}$ (230 K) < Frozen H_2O film, 80 K > 5×10^{12}	Cu, d = 17.5 cm, $\ell = 6$ m	$v < 5.6$ m/s	^{6}LiOH on ZnS 14 cm² <30 cm²>	1.4 <4> $(v < 5.6$ m/s$)$	0.07 <0.2>	/1.9/ </2.27/>
Vertical guide tube, Leningrad Inst. of Nucl. Phys.	Beryllium 420 K $(2-4) \times 10^{13}$	Stainless steel and ^{58}Ni-coated glass (7x6 cm²) $\ell = 8.5$ m	$v < 20$ m/s	^{3}He 42 cm²	36 $(v < 6.7$ m/s$)$	0.10	/3.9/
Neutron turbine, FRM, Munich	No converter 2×10^{13}	Feeding guide: Ni-coated glass 3.4×15 cm² $\ell = 11$ m, UCN beam: 10x20 cm²	$v < 20$ m/s (at the turbine exit)	BF_3 13 cm³	2.8^b $(v < 5.6$ m/s$)$	0.04	/2.21/^{c}

Neutron turbine, Inst. of Nucl. Research and Nucl. Energy, Sofia	No converter 2.4×10^{12}	Feeding guide: Stainless steel a = 4 cm ℓ ≈ 15 m UCN guide: Ni-coated steel	v < 6.2 m/s (at the turbine exit)	^{3}He 40 cm^2	0.1 (v < 6.2 m/s)	0.008	/2.23, 3.43/
Horiz. Channel, Research Inst. of Atomic Reactors, Dimitrovgrad	Zirconium hydride ≈ 300 K $(2-4) \times 10^{14}$	Stainless steel and Cu d = 7 - 10 cm ℓ = 9 m	v < 5.6 m/s	^{3}He 60 cm^2	11 (v < 5.6 m/s)	0.009	/3.8/

a A considerable deterioration of the copper guide tubes (but not of nickel or stainless steel guides) was observed in the course of several months, apparently due to a strong radiation corrosion of Cu.

b These values were determined by time of flight. In contrast to this direct method the technique of thin-foil transmission used in the other experiments tends to yields too optimistic UCN rates, especially in the presence of a large fraction of super-threshold neutrons. This is due to the fact that the foil not only blocks the sub-threshold neutrons, but also part of the faster neutrons by reflection. E.g., the foil value exceeds the TOF value for UCN at the FRM turbine by a factor of 2.

c The data apply to the FRM turbine with a partly reconstructed feeding guide tube which gave a four times higher intensity than the glass guide used previously /2.21/.

intermediary between interatomic distances in condensed matter and the resolving power of optical microscopes.

The extremely low energies under consideration can be utilized for studying low-energy excitations or slow diffusive processes in condensed matter, e.g. hyperfine interactions or critical fluctuations. Such investigations are in preparation in Munich. Previous similar work with cold neutrons in the meV-region has provided valuable insight into dynamical properties of H_2O in the solid, liquid and gas phase /4.2, 3/, and especially into hindered rotations in hydrogenous ʲliquids (e.g./4.4/).

4.1 Reaction and Inelastic Scattering Cross Sections

A complication of the 1/v law for the reaction cross sections arises at the lowest energies from the coherent elastic scattering of long-wavelength radiation in condensed systems, as described by the index of refraction. In consequence of refraction the group velocity is changed from v in vacuum to v' in the medium. It has been demonstrated both experimentally /2.24/ and theoretically within the framework of multiple scattering theory (for absorption: LENK /4.5/, for phonon scattering: IGNA-TOVICH /4.6, 7/) that the proper modification should consist in a replacement of 1/v by 1/v', as is also plausible from the classical argument that in the presence of refraction the interaction time with a target atom is changed by a factor v/v'.

It is well known that a long-range electromagnetic interaction between the scattering particle and the nucleus, in addition to the short-range nuclear forces, may modify the reaction cross section due to a change of the incident wave in the reaction zone (see, e.g. /4.1/). STEYERL has investigated the influence of the interaction of the magnetic moments of the neutron and the nucleus on the absorption cross section /4.8/. The conclusion of this analysis was that even in the presence of a strong hyperfine field a relative change of the capture cross section by only $\approx 10^{-10}$ can be expected. The effect of the atomic magnetic moment in a ferromagnet may be stronger by several orders of magnitude due to the larger magnetic moment of the electron and the large exchange field, but it will still be small. The experimental data discussed below do not contradict these conclusions.

4.1.1 Nuclear Capture

The measurement of total cross sections by low-energy neutron transmission experiments offers an absolute, accurate and rather direct method to determine cross sections for (n,γ), (n,α) or other slow-neutron induced nuclear reactions. The precise knowledge of such cross sections is often necessary for activation analysis, neutron flux monitoring and other purposes. Very slow neutrons may be used with advantage for such measurements, because the capture cross section is much higher than at thermal energies due to its 1/v variation, and also because the interference of Bragg scattering which is sample dependent may be avoided below the Bragg cutoff.

The more conventional methods of neutron spectroscopy, using mechanical monochromators with helical slots, could be used for cross section measurements down to the neutron energy range of ≈ 0.1 meV. For example, TODIREANU and CIOCA measured the total cross section of V and Cu in this way /4.9/. PALMGREN reported on a measurement on Cd at ≈ 8 μeV /3.18/, using the moving sample technique /3.17/. The energy range could be extended significantly by the vertical guide tube spectrometer for VCN at Munich /3.1/. STEYERL and VONACH determined at this spectrometer total cross sections of Au, Al, Cu, glass, mica and air down to the μeV range /2.24, 1.1/. This data clearly demonstrated the $1/v'$ variation of the capture cross section. Measurements with high precision (10^{-2} to 10^{-3}) were performed on Au, Co, Rh, Sc, V and Cu by DILG and MANNHART /2.26/ at the same spectrometer. As an example, Fig.4.1 shows the results of this experiment.

4.1.2 Neutron-Phonon Scattering

In most substances with a moderate or high capture cross section the thermal inelastic scattering contributes only slightly to the total cross section at room temperature. PLACZEK has shown that the phonon scattering may be accounted for in fair approximation by assuming single-phonon scattering, adopting the Debye model for the lattice vibrations and applying the incoherent approximation where also the coherent scattering is treated as if it were incoherent /4.10/. At neutron energies much below the characteristic excitation energy $k_B\Theta$ (Θ - Debye temperature) the phonon creation processes may be neglected compared to phonon absorption. In this limit the phonon scattering cross section may be represented (see, e.g. /4.11/) by the series

$$\sigma_{ph} = 3\frac{m}{N}(S + s)(k_B\Theta/E)^{1/2} \sum_{n=0}^{\infty} \frac{B_n}{n!(n+\frac{5}{2})} (\Theta/T)^{n-1}$$

$$= (S+s)\frac{m}{M}(k_B\Theta/E)^{1/2} \left[\frac{6}{5}\frac{T}{\Theta} - \frac{3}{7} + \frac{1}{18}\frac{\Theta}{T} - \frac{1}{1560}(\frac{\Theta}{T})^3 + \frac{1}{85680}(\frac{\Theta}{T})^5 - \frac{1}{4233600}(\frac{\Theta}{T})^7 + - \ldots\right]$$

(4.1)

which converges for $(T/\Theta) > 1/2\pi$. B_n are the Bernoulli numbers, T is the temperature, S+s the sum of the coherent and incoherent scattering cross section for the bound atom, and m and M are respectively the mass of the neutron and atom. Thus σ_{ph} varies $\sim 1/v$ (neglecting refraction), and $\sim$ T at higher temperatures. A different expansion shows that at low temperature σ_{ph} vanishes like $T^{7/2}$ /4.12/. BINDER has calculated that the additional "interference part" of the cross section, which varies also like $1/v$, amounts to ≈ 15 % in Al /4.11/.

Since the capture cross section of Al is small the inelastic scattering may be easily separated from the absorption which may be determined at low temperature where the phonon scattering vanishes. Fig.4.2 shows the experimental total cross section σ_T of Al plotted versus v' (and equivalent energy, wavelength and temperature scales) for various sample temperature /2.24/. These experiments and similar measurements on even weaker absorbers like Mg, Be and monocrystalline pyrolytic graphite /3.10/,

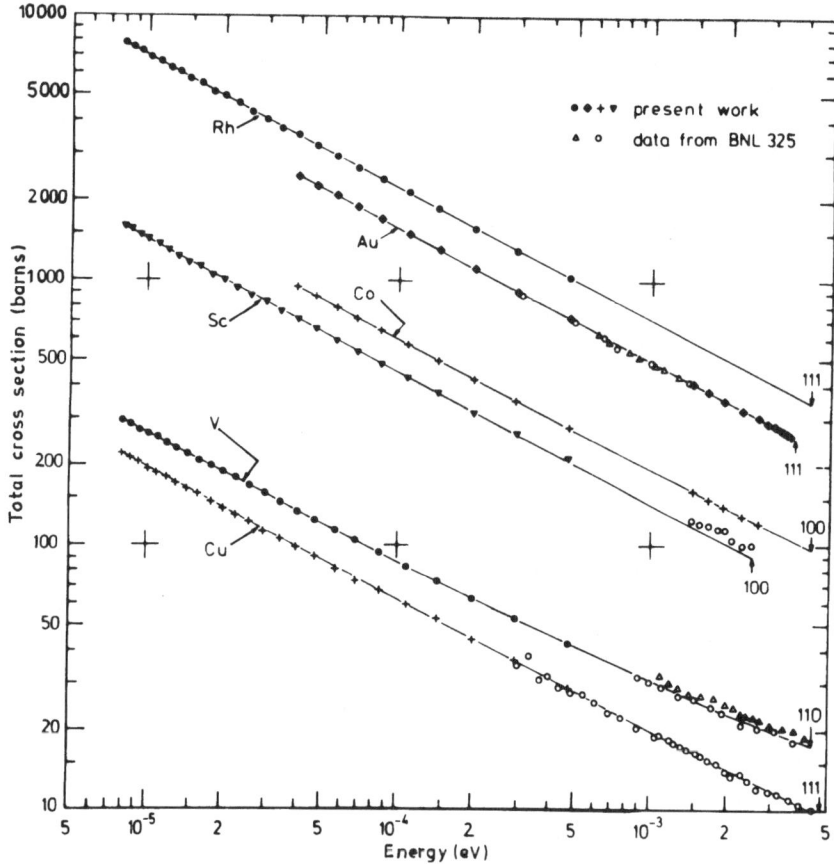

Fig.4.1. Total cross section of various substances for very low energy neutrons measured by DILG and MANNHART /2.26/

demonstrate that the phonon scattering which is determined as the difference between the cross sections at higher and at low temperature, follows the modified 1/v' law just like absorption. The experimental magnitude of σ_{ph} agrees with theory.

4.1.3 Scattering by Low Energy Excitations

While the 1/v law for the total inelastic-scattering cross section holds for neutron energies well below the characteristic excitation energies, low-energy modes with energies comparable to the neutron energy will give rise to characteristic structures both in the differential and integral scattering cross sections.

It has been shown theoretically that total cross section measurements with very cold neutrons could provide valuable information on hyperfine splittings /4.13 - 15/.

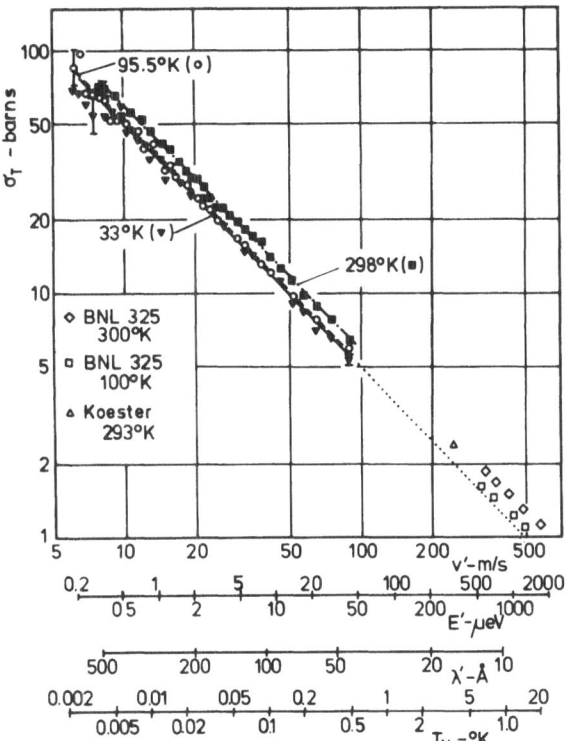

Fig.4.2. Total cross section
of aluminium at different
temperatures /2.24/

In addition, BINDER has suggested study in this way on low energies of magnetocrys-
talline anisotropy, low-lying rotational states in heavy molecules, and the slow
critical fluctuations in the magnetic phase transition just above the Curie point
/4.11, 12, 16, 17/. Another possible application is the investigation of the slow
tunneling motions between quasi-equilibrium atomic sites separated by a potential
barrier, e.g. in hindered rotations in molecular solids /4.18/. None of these ideas
has, however, so far been tested experimentally, mainly because the high resolution
of competing techniques is sufficient for a large number of such investigations.
Among these methods are the Mössbauer spectroscopy, the method of thermal neutron
back-scattering (at a Bragg angle of 90^0 /4.19/), and, since very recently, the
possibility of precise measurements of small-energy transfers by using the neutron
spin precession as a clock in the so-called spin-echo spectrometer /4.20/.

Recent experiments at the back-scattering spectrometers in Grenoble and Jülich
have shown, however, that there obviously exist hyperfine splittings of interest
and other low-energy states which lie significantly below the attainable resolution
width of $\approx 10^{-7}$ eV. The spectroscopy with very cold neutrons would have the definite
advantage that apparently no limit for the resolution exists, aside from intensity

considerations. Indeed, the theoretical intensity attainable from a Maxwellian spectrum in a double-differential scattering experiment is expected to increase with neutron wavelength (~ λ, approximately), for given resolution in momentum and energy transfer /2.28/. Therefore, a very high resolution spectrometer for ultracold neutrons, based on the use of gravity for monochromatization and energy analysis, is in preparation in Munich. It is expected to allow a resolution of $\approx 10^{-8}$ eV.

4.2 Elastic Scattering on Inhomogeneities

It is well known that waves are scattered by spatial fluctuations of the index of refraction. In X-ray or neutron small-angle scattering the scattering distribution is analysed and interpreted with respect to certain characteristics of the scattering zones. For example, it is possible to deduce from such experiments various size parameters for the partices, their number per unit volume, and sometimes information on their internal structure through the scattering form factor. A great number of specific applications of neutron small-angle scattering have been investigated in recent years, taking advantage of the low absorption and the magnetic interaction of neutrons. Thus, it is possible to study bulk material rather than thin films, as in electron microscopy, and also magnetic structures, in contrast to X-ray scattering. SCHMATZ et al. have recently presented a review of neutron small-angle scattering /2.18/.

Attempts have been made to use very cold neutrons for such purposes. The advantage is greater simplicity of the experiment, by virtue of the much wider angular scattering range of long-wavelength radiation and a strongly increased total scattering probability. Thus, it is even possible to perform simple transmission experiments instead of scanning the angular scattering distribution, since the variation of the total scattering cross section $\sigma(\vec{k})$ as a function of the incident-neutron wavevector $\vec{k}$ in principle contains the same information on the scattering objects as the differential scattering distribution as a function of momentum transfer $\hbar\vec{\kappa}$, $d\sigma/d\Omega = |f(\vec{\kappa})|^2$. This is plausible for random orientation of scattering zones where the dependence on the direction of $\vec{k}$ and $\vec{\kappa}$ disappears. In this case it is easy to see that the differential scattering law may in principle be retrieved from the total scattering cross section

$$\sigma(k) = \frac{2\pi}{k^2} \int_0^{2K} \kappa |f(\kappa)|^2 d\kappa \qquad (4.2)$$

by differentiation. In practice, no differentiation is necessary, as the relevant characteristic parameters may be obtained from $\sigma(k)$ in an equally direct way as from $|f(\kappa)|^2$ /4.21, 2.15/.

It has been shown that the method of VCN cross section measurements yields reliable results for well-known systems like the Guinier-Preston zones in AlZn alloys /2.15/ or an aqueous colloidal dispersion of spherical silica particles /4.22/. LERMER and STEYERL /2.17/ and LENGSFELD /4.22/ have studied the magnetic domains and domain walls in various ferromagnets in this way. Fig.4.3 shows the results of /2.17/ for the total cross section σ_T of polycrystalline nickel for various values of the applied magnetic field. The data at low magnetic field strengths exhibit a significant deviation from 1/v, while nearly pure 1/v behaviour is reached at a saturation field of 100 Oe. Thus, it is concluded that the deviation from 1/v is caused by strong scattering from the orientational variations of magnetization associated with the domain structure. A theory was developed which allows one to interpret the scattering effect with respect to the mean domain size and domain wall thickness /4.21, 2.17/. The values obtained for the domain walls in cobalt agree with theoretical predictions, while some discrepancy by factors of 2 to 3 was observed for iron and nickel. These data are of interest in view of the great difficulties and widely scattering results of any other existing technique to measure the thickness of domain walls (see, e.g. /4.23/).

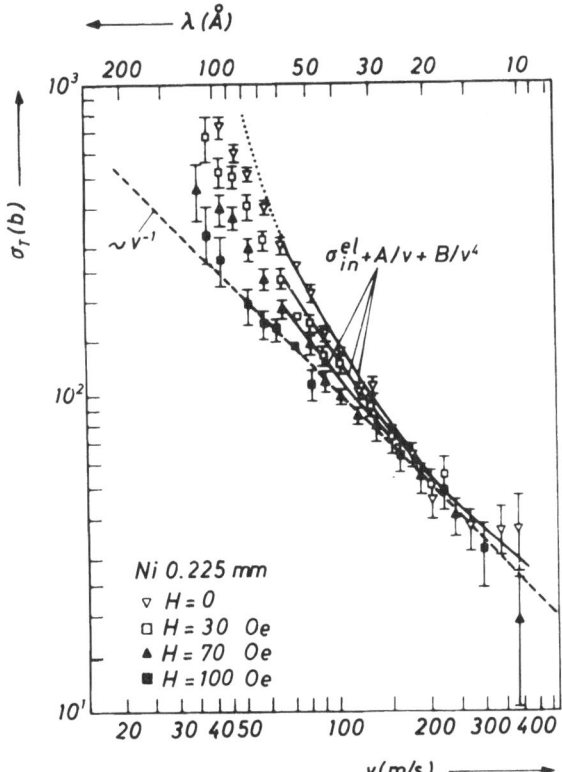

Fig.4.3. Total cross section of nickel for various values of the applied magnetic field /2.17/

As a possible further application, BINDER has calculated the total scattering cross section of very cold neutrons by the vortex lattice in type II superconductors /4.16/.

4.3 Transmission Through Thin Films

Neutrons incident from vacuum on a thin slab of material are subject to the action of the scattering potential U. If U is positive and the foil much thicker than the neutron wavelength, neutrons with energies E < U will not be able to penetrate the potential barrier but will be totally reflected. For E not much greater than U, reflection is still significant, whereas for E >> U practically all neutrons will be transmitted. Consequently, we expect the energy dependence of transmission (or reflection) to be given by smoothed step functions.

This transmission behaviour has been demonstrated experimentally with very cold neutrons in Munich /4.24, 3.25, 3.10, 2.21/. The measurements allow one to determine the scattering potential, and hence, the foil density or atomic scattering length, from the position of the steep edge. Some information can also be gained on impurities like the H_2O content which has a large effect on the scattering potential.

Attempts have been made to observe the interference pattern in very thin films of thickness comparable to the neutron wavelength, where the interference of partial reflection and transmission of the neutron wave at the two surfaces is important /4.24/. The expected structure was observed for 450 to 800 Å thick carbon and gold films. It was noted that such measurements could be applied to determine absolute values for the film thickness and density since the interference pattern is sensitively affected by these quantities.

ANTONOV et al. have analysed theoretically the reflection and transmission properties of an interference filter for ultracold neutrons consisting of a periodic system of alternating layers of two substances with strongly different scattering potentials /4.25/. They concluded that the Bloch-wave-type band structure in the reflection and transmission spectrum could be utilized for high-resolution spectroscopy with neutrons of energy $10^{-8} - 10^{-4}$ eV.

In ferromagnets there are two distinct values for U, due to the contribution of the magnetic interaction potential which is positive or negative depending on the configuration of the neutron spin relative to magnetization (see (2.8)). Consequently, the transmission curve of unpolarized neutrons through a ferromagnetic film may exhibit two edges, as illustrated for nickel by the experimental data of Fig.4.4 /2.21/.

It has been known for a long time that the spin dependence of the scattering potential may be utilized for neutron polarization and polarization analysis. For thermal or cold neutrons it is usual to use the reflection geometry for such purposes. On the other hand, the transmission geometry is more suitable for ultracold

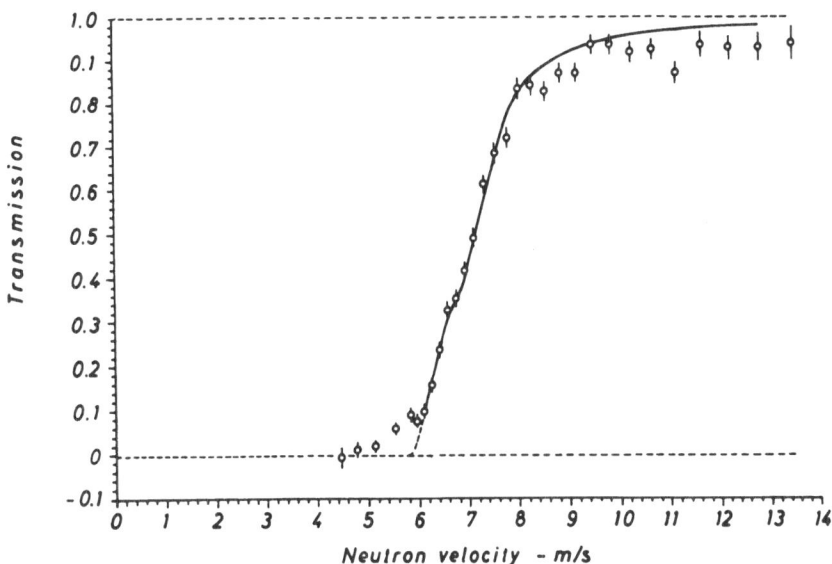

Fig.4.4. Perpendicular transmission of very cold neutrons through an unmagnetized nickel film (1500 Å, evaporated on aluminium foil). The data shows two steep edges which are determined by the different scattering potentials for neutron spin parallel and antiparallel to domain magnetization /2.21/

neutrons because it permits one to work with a wide angular beam divergence. The experience gained so far in UCN polarization using ferromagnetic foil transmission will be discussed in Section 6.1.3.1.

POKOTILOVSKY et al. have proposed to use ferromagnetic foils as fast gates for polarized UCN /4.26/. Such devices might be useful if pulsed neutron sources were to be employed as UCN sources, as proposed by SHAPIRO /1.9/. Theoretically, a high density of UCN could be accumulated in a neutron container filled with neutrons only during the short neutron pulse cycles. The practical feasibility of this conception, however, relies critically both on fast shutter operation and on long containment times for polarized neutrons.

5. Neutron Containment in Material Bottles

Considering that neutrons ordinarily interact with matter very weakly it may at first appear like a "Schildbürgerstreich" to try to catch a neutron in a box. (The citizens of Schilda are said to have attempted to trap light in various containers for the purpose of illuminating their triangular city hall, where they had forgotten to install windows). Nevertheless, ZEL'DOVICH has pointed out in 1959 that neutrons of sufficiently low energy should experience total external reflection from suitable materials at any angle of incidence, and hence, it should be possible to trap neutrons in closed cavities where they keep bouncing to and fro between the walls /1.3/. In the process of total reflection the neutron wave penetrates into the medium only a depth of roughly $1/k_{cr} \approx 100$ Å, where k_{cr} is the limiting wavenumber for total reflection at normal incidence. Therefore, the neutron loss per reflection due to absorption into a nucleus or inelastic scattering out of the energy range suitable for storage, should be small. For weakly absorbing substances like graphite and bottle dimensions of ≈ 10 cm the reflection losses should be even negligible as compared to the neutron β-decay rate with a lifetime of about 1000 s.

GUREVICH and NEMIROVSKY treated as a different principle of neutron reflection the "metallic" reflection which is based on the high reflectivity of strongly absorbing media, as in light optics /5.1/. They came to the conclusion that even for the strongest neutron absorber known (^{157}Gd) this reflection mechanism is much less effective than total reflection.

As a kind of curiosity, we may mention also a paper of FOLDY of 1966 /5.2/ in which he proposed a neutron bottle using superfluid helium at 10^{-5} K as the wall material. He calculated that about 10^{15} neutrons/cm^3 could be stored under the assumption of a completely degenerate neutron Fermi gas up to the wall scattering potential. Nuclear and high-energy physicists would be delighted at a free neutron target of such a high density, but unfortunately Foldy did not tell us how the bottle could be filled up to this theoretical limit.

In 1968 SHAPIRO pointed out that the use of stored ultracold neutrons could lead to a significant improvement of sensitivity in the very fundamental experiment aimed at a search for an electric dipole moment of the neutron /1.5/. The underlying ideas and the principle of the proposed experiment will be discussed in Section 6.1. The thus increased interest in neutron storage stimulated a research group of Professor Shapiro in Dubna and at the Institute of Atomic Energy in Moscow to investigate this problem experimentally. They reported in 1971 on the first successful storage of UCN in closed vessels made of various materials: copper, pyrolytic graphite, beryllium and teflon /2.12/.

In the meantime similar experiments have been performed many times in various laboratories. The surprising result of all these endeavours seems to be that the neutrons are observed to disappear from the bottle at a much faster rate than expected.

5.1 The Elementary Theory

We consider the process of total reflection of ultracold neutrons with energy E < U incident from vacuum on the ideally plane and clean surface of a semi-infinite medium characterized by the scattering potential U. It is well known that the various aspects of wave attenuation may be described phenomenologically by introducing an imaginary part $-U_1$ into the potential, i.e. by replacing the medium by a complex potential

$$V = U - iU_1. \tag{5.1}$$

According to (2.7) and (2.3) U is related to the real part of the scattering amplitude, b_r, by

$$U = 2\pi\hbar^2 Nb_r/m. \tag{5.2}$$

There has been much discussion in the literature as to whether it is adequate to represent the imaginary part $-U_1$ in an analogous way by the imaginary part b_i of the scattering amplitude, which is related to the total cross section σ_T by the optical theorem

$$\sigma_T = -4\pi b_i/k. \tag{5.3}$$

In a crystalline medium the coherent elastic scattering cross section σ_{el} does not lead to beam attenuation below the Bragg limit. This is consistent with the analyses of multiple scattering by SOLBRIG /5.3/ and by IGNATOVICH and LUSHCHIKOV /2.7/ which show that in this case the scattering amplitude for the isolated atom, b, should be replaced by the effective crystalline scattering amplitude $\frac{b}{1 - ik_ob} \equiv b + ik_o\sigma_{el}/4\pi$ where k_o is the neutron wavenumber in vacuum. There is some controversy regarding the effect of σ_{el} in an ideally random medium. According to standard theory /2.6/ the appropriate incoherent contribution of σ_{el} to U_1 should be

$$\hbar^2 Nk_o\sigma_{el}/2m. \tag{5.4}$$

On the other hand, in /4.5/ LENK finds a dependence not on k_o but on k', the real part of the wavenumber in the medium, which is virtually zero under the condition of total reflection. In any event, it has been shown by IGNATOVICH /2.6/ that any incoherent-elastic scattering will not give rise to reflection losses for ultracold neutrons but only to some nonspecularity, because the scattered wave cannot penetrate the medium and must, therefore, return to the vacuum.

For this reason, it seems appropriate to consider in U_1 only the cross sections for nuclear capture, σ_c, and inelastic scattering, σ_{ie}, i.e.

$$U_1 = -2\pi\hbar^2 N(b_i)_{eff}/m = \hbar^2 Nk(\sigma_c+\sigma_{ie})/2m. \tag{5.5}$$

U_1 is constant at low neutron energies due to the 1/k variation of both σ_c and σ_{ie}.

For the vast majority of substances, $U_1 \ll U$. In this limit the probability of neutron loss at a wall collision is given by

$$\mu = 2\eta \frac{x}{(1-x^2)^{1/2}} \qquad (5.6)$$

where $\eta = U_1/U$, and $x = (v_0\cos\theta)/v_{cr}$ is the ratio of the vacuum velocity component normal to the wall to the limiting velocity for total reflection at normal incidence, $v_{cr} = \hbar k_{cr}/m = (\hbar/m)(4\pi Nb_r)^{1/2}$. As $x \to 1$, μ diverges, in consequence of the divergence of the penetration depth $\delta = (1/k_{cr})(1-x^2)^{-1/2}$.

For an isotropic angular distribution of incident neutrons the average absorption probability per reflection is calculated to be /1.9/

$$\bar{\mu} = (2\eta/y^2)(\arcsin y - y\sqrt{1-y^2}), \qquad (5.7)$$

where $y^2 = v_0^2/v_{cr}^2 = E/U$. This function is plotted in Fig.5.1.

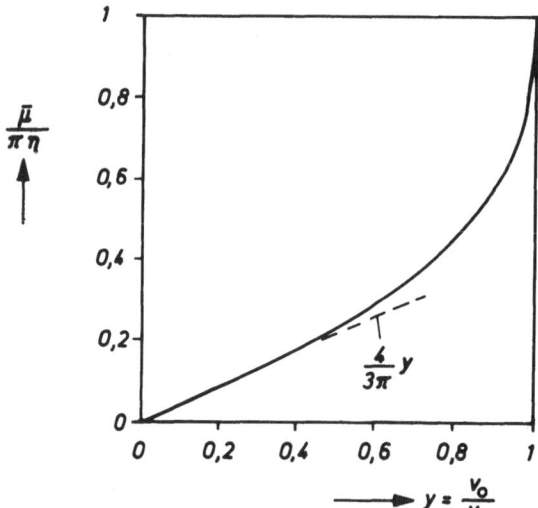

Fig.5.1. Theoretical velocity dependence of $\bar{\mu}/\pi\eta$, the normalized loss probability for ultracold neutron reflection, averaged over the angle of incidence on the wall

For monoenergetic neutrons contained in a bottle the loss probability per s should be

$$W = 1/T = v_0\bar{\mu}/\ell + 1/\tau \qquad (5.8)$$

where $\bar{l}$ is the mean free path between successive wall collisions and $\tau \cong 1000$ s is the lifetime for the free neutron β-decay. T is the containment time. The gaskinetic appraoch yields $\bar{l} = 4V/S$, where V and S are respectively the volume and the surface area of the bottle. Similarly, the time constant for filling (or emptying) a loss-less bottle is calculated to be $T_f = 4V/Av_0$, where A is the area of the aperture.

If the bottle is filled with neutrons of different velocities, the time dependencies both of neutron survival in the bottle and of the filling behaviour are expected to be nonexponential because W and T_f depend on v_0.

Additional complications may arise from the inhomogeneous spatial distribution in a large bottle due to gravity. This effect has been analysed by IGNATOVICH and TEREKHOV /5.4/. Furthermore, in bottles with high geometrical symmetry and mirror reflecting walls the angular flux distribution may be far from isotropic. IGNATO-VICH has presented analytical solutions of this problem in the case of certain simple geometries /5.5/.

5.2 Experiments and Results

Fig.5.2 shows the scheme of the arrangement used in most of the storage experiments performed at the Institute of Atomic Energy, Moscow /1.9, 3.13/, and at the Research Institute of Atomic Reactors, Dimitrovgrad /3.8/. The storage chamber 1 consists of a 0.2 to 3 m long tube with a diamter of 6 to 14 cm. The vessel is repeatedly filled with neutrons through valve 2. Then the neutrons are kept in the closed system for a variable storage time t. Subsequently, valve 3 is opened and the surviving neutrons discharge to the detector 4. Tight fitting of the copper valves was ensured by the use of pneumatic drives.

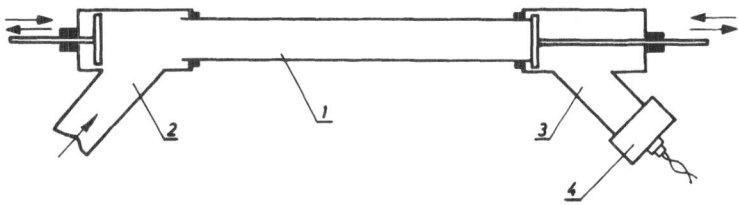

Fig.5.2. Scheme of the experimental arrangement for ultracold neutron storage measurements performed at the Institute of Atomic Energy, Moscow /3.13/. 1 exchangeable storage vessel in the form of a tube; 2, 3 admission and release channels connected to the storage volume by pneumatically driven valves; 4 detector

In some experiments the storage section was installed in a rotary elbow connected to a horizontal UCN channel. With such an arrangement the spectral range of stored neutrons could be controlled by varying the bottle elevation above or below the guide-tube level. Similarly, the emergent neutrons were allowed to accelerate by falling through a vertical distance between the storage vessel and the detector, in order to improve the transparency of the detector window and hence the detector efficiency.

Arrangements of the type of Fig.5.2 were also used in the experiments at the Institute of Nuclear Physics, Leningrad /3.9/ and at the Universities Research Reactor, Risley /3.42/.

A somewhat different scheme was applied in the experiments at the FRM, Munich /3.10/. Fig.5.3 shows the arrangement of the investigated pyrolytic graphite bottle which was connected to the vertical guide tube for UCN. In this experiment the shutters were made of the same material as the bottle body. The shutters move in slides leaving very small gaps. The low-energy spectral cutoff of the detected stored neutrons could be varied by placing different thin transmission foils with appropriate scattering potentials in front of the detector.

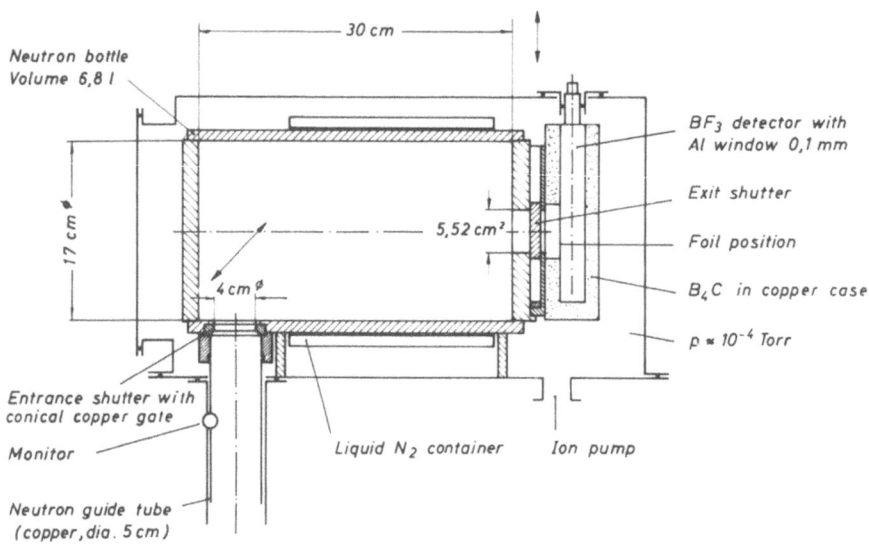

Fig.5.3. Scheme of the graphite bottle investigated at the reactor FRM, Munich /3.10/. The graphite exit and entrance shutter moves respectively in and perpendicular to the drawing plane

In most experiments the neutron efflux was time analysed and found to be nearly exponential. Containment lifetimes were determined by measurements of the integral count rate of surviving neutrons after variable durations of the storage cycle. Fig.5.4 shows typical results as determined for a pyrolytic graphite bottle at room temperature /3.10/. The data points represent different velocity intervals of bottled neutrons, ranging from the indicated cutoff values v_ℓ of the various transmission foils used to the critical velocity of 5.8 m/s for graphite. The resulting containment lifetimes T (and sometimes the variation of T with time due to the non-exponential behaviour expected and observed for wide spectral ranges) are indicated.

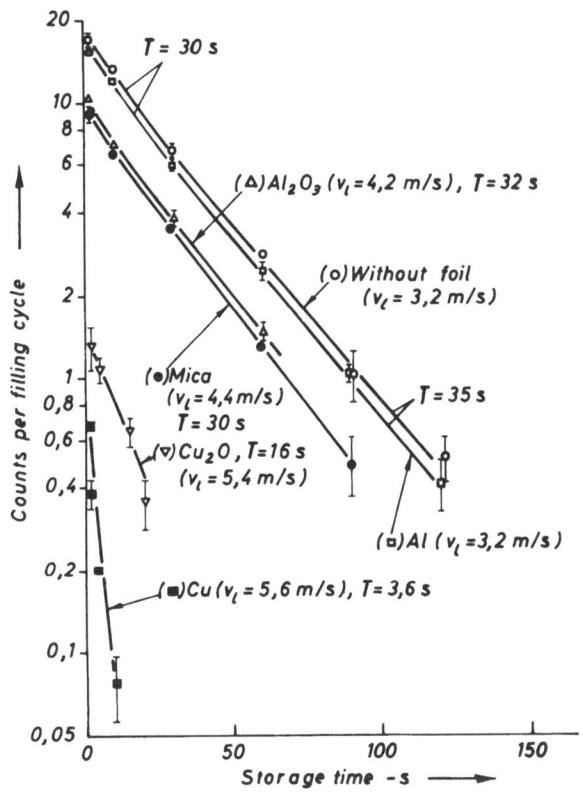

Fig.5.4. Results of measurements of containment lifetimes T in a graphite bottle at room temperature /3.10/. Various spectral ranges of detected neutrons were provided by suitable transmission foils placed in front of the detector. The foils determine the lower spectral cutoff by their limiting transmission velocity v_ℓ. The upper spectral limit is at 5.8 m/s for the pyrolytic graphite used

The mean loss probability per wall collision calculated from the data for the full spectrum was reported to be about 8×10^{-4} which is as much as 160 times higher than the theoretical value. Furthermore, the increase of reflection loss near the critical velocity for graphite was observed to be steeper than for the theoretical curve of Fig.5.1.

An improved spectral resolution was achieved by the gravity spectrometer used in Moscow. The results of GROSHEV et al. /5.6/ for the differential containment life-times in a copper trap are represented in Fig.5.5 in the form of the experimental absorption probability $\bar{\mu}$ per wall collision plotted versus the neutron velocity. The reported magnitude of $\bar{\mu}$ is nearly the same as for the graphite bottle, although the theoretical loss probability for Cu is as much as 45 times higher than for graphite due to the much larger cross section for nuclear capture. The data show the same qualitative v-dependence as the theoretical curve 1 in Fig.5.5. The hypotheses lead-ing to the curves 2, 3 and 4 in Fig.5.5 will be discussed in Section 5.3.2.

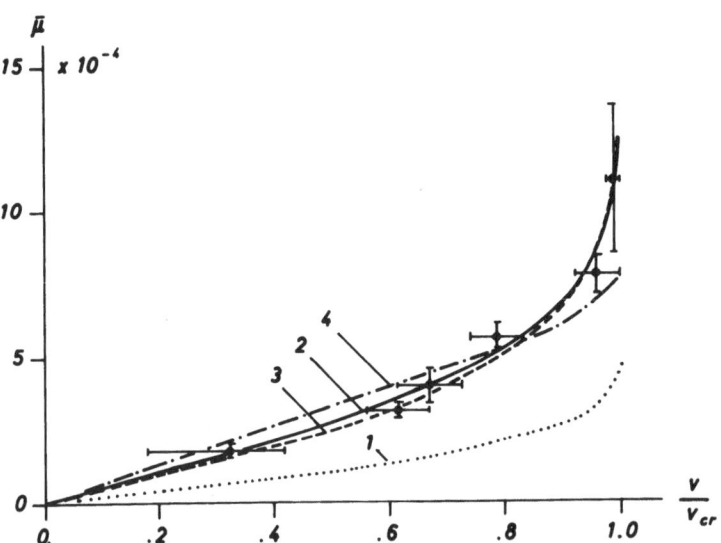

Fig.5.5. Absorption coefficient for ultracold neutrons in a copper trap plotted versus the neutron velocity /5.6/. The spectral ranges for the experimental points (circles) are indicated by horizontal bars. 1 elementary theory; 2 elementary theory but with an adjusted, 2.6 times enhanced, $\bar{\mu}$-value; 3 model of a soft potential step for the wall with an adjusted width of the smooth edge (110 Å); 4 hydrogenous contamination on the wall with an adjusted thickness (55 Å)

The measurements discussed above are examples for a large number of similar experiments on neutron storage in material traps, which have been performed in several laboratories since the pioneering work of Shapiro's group in 1971 /2.21/. A survey over these efforts and the main results obtained is presented in Table 5.1. Inspection of these data shows that the discrepancy between the experimental and expected containment lifetimes still persists. It seems curious that about the same loss probability per reflection of $\approx 10^{-3}$ or slightly below is obtained with substances with widely different capture cross sections. The only agreement between experiment and theory was observed with borated glass where the loss probability is exceptionally high due to the $^{10}B(n,\alpha)^{7}Li$ reaction.

A number of experiments were carried out in order to examine the influence of specific experimental parameters on the containment lifetime:

a) The effect of gaps was checked by providing tighter fits or by control experiments with wider gaps or absorbers simulating gaps. The observed probability of neutron loss in a gap is consistent with the proportion of gap area to the entire bottle surface, and may be neglected in most cases.

b) Any sizable effect of neutron heating by collisions with gas molecules could be excluded by experiments which showed that a variation of pressure below, say, 10^{-3} mm Hg does not affect the containment time, in agreement with the exception based on the known cross section data (e.g., /3.8/).

c) Some apparent improvements in lifetime, especially in long cylindrical glass and quartz tubes, could be attributed to the long survival of neutrons which experience mirror reflections at low angles to the axis. The mean free path for such neutrons is much larger than the gaskinetic value, and hence the $\bar{\mu}$-values listed in Table 5.1, which were calculated on the basis of gaskinetics, are overoptimistic. The indicated control experiments with short tubes indeed yielded considerably higher loss probabilites /5.7/.

d) In the absence of such geometrical effects it could be demonstrated by a variation of the trap dimensions that the loss rate per sec is proportional to the frequency of wall collisions, and hence the reflection process is the most likely source of neutron loss /3.8/.

e) Accordingly, special attention was paid to clean surface conditions. SHAPIRO reports that a factor of 2 in lifetime was gained both for Cu and glass by improved surface treatment /1.9/, using hydrofluoric acid for glass instead of alcohol, and the electrolyte H_3PO_4 for Cu rather than $H_3PO_4 + CrO_3$. Similarly, a factor of 3 was gained with pyrolytic graphite by outgassing at $400^{\circ}C$ in 10^{-6} Torr /3.10/, as compared to previous data /2.12/.

f) It was hoped that a variation of the wall temperature in storage experiments could provide a clue for understanding the high reflection losses. On the one side, heating is expected to remove organic, hydrogenous contaminations which could cause

Table 5.1. Results of various experiments on neutron containment in material traps

Institute	Bottle material and dimensions d - diameter L - length	Treatment of the internal surface, Measuring temperature T (K)	Neutron velocity interval in m/s	Mean containment lifetime in s	Loss probability per reflection $<\bar{\mu}>$		Reference
					Experiment	Expectation from cross section data	
Inst. of Atomic Energy, Moscow	d = 14 cm L = 1.74 m	Cu T = 300	3.2 - 5.6	33	9×10^{-4}	2.4×10^{-4}	
		Cu T = 520	"	same	same	"	/2.12/
	Pyrolytic graphite	T = 300	3.2 - 5.6	11	2.7×10^{-3}	5.2×10^{-6}	
	"	T = 100	"	same	same	5.2×10^{-7}	
	Be	T = 300	"	$\approx$ same	$\approx$ same	7.2×10^{-6}	
	Teflon $(CF_2)_n$	"	3.2 - 4.8	$\approx$ same	$\approx$ same	$\approx 10^{-5}$	
	Cu d = 8.5 cm L = 2 m	Electropolished in H_3PO_4 T = 300	3.2 - 5.6	30 $\pm$ 2	6.0×10^{-4}	2.4×10^{-4}	/1.9/[a]
	"	T = 670	"	"	"	2.5×10^{-4}	/5.7/
	"	T = 300	1.5 - 2.9	105 $\pm$ 7	3.1×10^{-4}	9.5×10^{-5}	/1.9/
	L = 0.2 m	"	3.2 - 5.6	22	6.7×10^{-4}	2.4×10^{-4}	/5.7/

Material	Treatment	T	Range	Value (±)			Ref.
Glass $d = 8$ cm $L = 3$ m	Repeated treatment in 3 % HF	$T = 300$	$3.2 - 4.2$	104 ± 7	2.0×10^{-4}	9.4×10^{-6}	/1.9/[a]
"		$T = 570$	"	"	"	1.1×10^{-5}	/5.7/
$L = 0.2$ m		$T = 300$	"	36	4.8×10^{-4}	9.4×10^{-6}	/5.7/
Teflon (CF_2) $d = 8$ cm	Treatment with hot NaOH solution	$T = 300$	$3.2 - 4.8$	10 ± 1.4	2×10^{-3}	$\approx 10^{-5}$	/1.9/
Borated glass	Rinsing with distilled water		$1.7 - 4.3$		3×10^{-3}	3×10^{-3}	/3.13/
Quartz $d = 6$ cm $L = 0.8$ m		$T = 300$	< 4.1	100 ± 5	1.4×10^{-4}	1.6×10^{-5}	/5.7/[a]
Stainless steel $d = 9$ cm $L = 1.8$ m	Electropolished	$T = 300$	< 5.6	25	7.6×10^{-4}	1.7×10^{-4}	/5.7/
FRM, Munich Pyrolytic graphite $d = 17$ cm $L = 0.3$ m	Baked at 400°C and 10^{-6} Torr	$T = 300$	$3.2 - 5.8$	32 ± 2	8.3×10^{-4}	5.2×10^{-6}	/3.10/
Cu $d = 17$ cm $L = 0.3$ m	Electroplated on graphite	$T = 300$	$3.2 - 5.6$	22 ± 2	1.2×10^{-3}	2.4×10^{-4}	/3.10/

Table 5.1 (continued)

Institute	Bottle material and dimensions d – diameter L – length	Treatment of the internal surface, Measuring temperature T (K)	Neutron velocity interval in m/s	Mean containment lifetime in s	Loss probability per reflection $\langle\bar\mu\rangle$		Reference
					Experiment	Expectation from cross section data	
Inst. of Nuclear Physics, Leningrad	Ni d = 7 cm L = 1.2 m " Glass	Electroplated on Cu T = 300 T = 470 or precedent heating T = 300	< 6.2 " < 4.2	15 ± 1 same 25 ± 3	6.7×10^{-4} same 6.7×10^{-4}	1.5×10^{-4} " 1.5×10^{-5}	/3.9/
Universities Research Reactor, Risley	Cu d = 19 cm L = 2.18 m Glass 15 x 5 cm^2 L = 2.18 m	Rinsing with alcohol T = 300 "	< 5.6 < 4.2	16 9	2.4×10^{-3} 2.1×10^{-3}	2.4×10^{-4} 1.5×10^{-5}	/3.42/

Research Inst. of Atomic Re-actors, Dimitrov-grad	Stainless Steel $d = 64$ cm $L = 1.05$ m	Electropolished $T = 300$	< 5.6	47 ± 3	2.3×10^{-3}	1.7×10^{-4}	/3.8/

a These data were taken with traps in the form of long cylindrical tubes. Thoeretical considerations and also the indicated control experiments with short tubes show that the relatively long containment lifetimes observed are due to the prepon-derance of mirror reflections from the tube wall for which the mean free path is much larger than calculated in the common diffuse-reflection approximation. Due account of this effect reduces the μ-values given to those obtained with short or corrugated tubes.

enhanced losses due to the large inelastic scattering cross section of hydrogen. On the other side, cooling should significantly reduce all inelastic scattering effects and therefore the reflection loss, especially for weak absorbers like graphite or quartz, where the probability of nuclear capture is greatly outweighed at room temperature by inelastic scattering processes. In most solids the thermal scattering cross section decreases by about one order of magnitude on cooling from 300 to 100 K. In view of these considerations it was surprising that in the experiments performed at Moscow /1.9, 2.12, 5.7/ and Leningrad /3.9/ the experimental containment lifetime was reported to remain constant within a few percent over a wide range of temperature, e.g. for glass between 100 and 770 K and for graphite between 100 and 300 K.

g) The hypothesis of the presence of H_2O on real surfaces in sufficient quantities to account for the room temperature data ($\approx$ 100 Å) is rendered unlikely also by the negative result of prolonged surface exposure to D_2O in order to exchange H_2O by D_2O which has a much smaller cross section for slow neutrons /5.7/.

h) The experiments with graphite as well as a measurement with BeO /3.9/ yielded about the same loss rates as substances of crystalline density. This indicates that the strong structural inhomogeneity and porosity in these materials are of no crucial consequence for the reflection properties. This experimental finding checks with the expectation that the neutron wave should average over density fluctuations of sufficiently small extensions as compared to the ultracold neutron wavelength of $\approx$ 1000 Å.

In addition to studying the containment times under various experimental conditions, attempts have been made to detect the neutrons disappearing from the bottle. Cursory measurements mentioned by GROSHEV et al. /2.12/ had indicated that as UCN travel through copper tubes there is no gradual heating which could arise, for example, from vibrations of the walls. Very recent experiments done at the Institute of Atomic Energy, Moscow, where a thin-walled copper trap was surrounded by an annular detector for very slow neutrons (5 < v ≳ 100 m/s), failed to detect any leakage of slightly heated neutrons through the wall /5.8/.

A different scheme of filling a neutron bottle was tried in Munich /3.10/ and at the Lebedev Institute, Moscow /5.9/. The trap was filled with UCN at the reactor active zone and then moved to an experimental site where the radiation background is low. The neutron losses due to this displacement are negligible provided that the accelerations are kept smaller than, say, the gravitational acceleration. The main difficulty in these experiments is the discrimination of UCN detection against the intense radiation from the activated structural material of the bottle. Consequently, although neutron containment could be demonstrated in these experiments, the results were inconclusive in regard to the containment time.

5.3 Theoretical Attempts to Explain the Data

The theoretical investigations into the possible reasons for the short experimental containment lifetimes are proceeding along two different lines: On the one side, the "elementary theory" has been modified in order to account for more realistic surfaces which are neither ideally plane nor pure. In this connection the processes leading to inelastic neutron scattering were carefully analysed, and in particular the role of acoustic vibrations and various other types of low-frequency oscillations. The other line of thought is concerned with a scrutiny of the conventional optical potential formalism for slow neutrons. Even the idea that we may be seeing a new, fundamental limit to the coherence length of the neutron wave was raised.

5.3.1 The Perturbation Method

The deviation from an ideal vacuum-medium boundary is frequently considered to be sufficiently small for a perturbation approach. A large class of surface perturbations may be treated by a distorted-wave Born approximation where the unperturbed wave is taken to be the wave field for the ideal boundary, consisting of the incident, the specularly reflected and the penetrating plane waves. The general formalism developed by IGNATOVICH /4.6, 7, 2.7/ and STEPANOV /5.10, 11/ starts from the Schrödinger equation describing the nuclear neutron interaction with the wall

$$\left[\frac{2mi}{\hbar} \frac{\partial}{\partial t} + \nabla^2 - u_0 \theta(z) - \delta u(\vec{r},t) \right] \Psi(\vec{r},t) = 0. \tag{5.9}$$

$u_0 = 4\pi Nb$ is the scattering potential of the hypothetical wall which is assumed to consist of fixed atoms with scattering length b and to occupy the half space $z > 0$. $\theta(z)$ is the step function (= 1 for $z > 0$ and = 0 for $z < 0$) and $\delta u(\vec{r},t)$ the potential of the perturbation which depends on time if dynamic processes are considered. In this general case the solution of (5.9) may be written, up to the second order in δu

$$\Psi = \Psi_0 + G \cdot \delta u \cdot \Psi_0 + G \cdot \delta u \cdot G \cdot \delta u \cdot \Psi_0 + \dots \tag{5.10}$$

where $G \cdot \delta u \cdot f$ symbolically stands for

$$G \cdot \delta u \cdot f = \int G(\vec{r},t;\vec{r}',t') \; \delta u(\vec{r}',t') f(\vec{r}',t') d\vec{r}' dt'.$$

$\Psi_0(\vec{r},t)$ is the unperturbed wave field for the step-potential. The Green function $G(\vec{r},t;\vec{r}',t')$ satisfying the equation

$$\left[\frac{2mi}{\hbar} \frac{\partial}{\partial t} + \nabla^2 - u_0 \theta(z) \right] G(\vec{r},t;\vec{r}',t') = \delta(\vec{r}-\vec{r}')\delta(t-t') \tag{5.11}$$

may be represented by the complete set of eigenfunctions of the homogeneous equation,

$$\begin{pmatrix} Y_{k_\perp}(z) \\ H_{k_\perp}(z) \end{pmatrix} \exp(i\vec{k}_{\shortparallel}\cdot\vec{r}_{\shortparallel} - i\omega t).$$

The subscript $_{\shortparallel}$ and $_\perp$ denotes respectively the component parallel and perpendicular to the wall. $k_\perp$ is subject to the condition $k_\perp^2 + k_{\shortparallel}^2 = 2m\omega/\hbar$. Even in the time-independent case where $\omega = \omega_0 = \hbar k_0^2/2m$, $k_{\shortparallel}$ must be allowed to assume any value, even larger than k_0, the incident neutron wavenumber. (The case $k_{\shortparallel} > k_0$ corresponds to the creation of exponentially damped waves in the z-direction which, however, must be considered in second-order calculations.) $Y_{k_\perp}(z)$ and $H_{k_\perp}(z)$ are linearly independent solutions of the time-independent one-dimensional wave equation, representing respectively waves incident from vacuum and from the medium. (Solutions with indefinite exponential growth in space must be included in this set, although they do not correspond to physically possible states.)

5.3.2 Considerations of Various Wall Perturbations

The formalism outlined in the preceding subsection has been applied to calculate loss probabilities in UCN wall collisions arising from the neutron interaction with bulk phonons and with various types of low-frequency oscillations: acoustic wall vibrations, surface waves and the slow motion of heavy clusters /4.6, 7/. The effect of static surface microroughness was studied in the same way /3.30, 5.10, 11, 2.7/. Different approaches had to be used to analyse imperfections of a larger scale, like macroscopic surface roughness /3.30/ and contamination layers /5.12/.

IGNATOVICH's analysis of neutron-phonon incoherent scattering processes /4.6, 7/ leads to exactly the same expression for the loss probability as the elementary theory with an imaginary part of the pseudopotential given by (5.5). This result as well as the cross section measurements discussed in Section 4.1.2 seem to rule out a speculation of LENK /5.13/ that in a refracting medium the inelastic scattering cross section might increase anomalously, namely $\sim v'^{-2}$ rather than $\sim v'^{-1}$, as the neutron velocity in the medium, v', vanishes.

The hypothesis of the significance of vibrating clusters was discussed by IGNATOVICH in /4.7/. The author comes to the conclusion that the theoretical loss probability due to neutron heating by clusters vibrating as a whole cannot explain the experimental data at room temperature, except under the following drastic assumptions: a) Virtually the whole surface volume consists of clusters (of dimensions 60 - 70 Å) and b) these clusters oscillate independently of each other like Einstein oscillators (with frequencies of $\approx 10^{10}$ cps). Although the cluster theory is appealing in view of a similar hypothesis to explain the specific heat anomaly in amorphous substances at low temperature, the above two conditions seem to be mutually exclusive since objects in close contact should be excited to collective rather than indepen-dent motions. For collective motion, on the other hand, the square amplitude of incoherent scattering on clusters, and hence the total scattering probability, should be strongly suppressed. (For independent oscillators it is determined by $\sigma_{inc} \sim n^2$, where n is the average number of atoms in a cluster, whereas for coupled oscillators of different size incoherent scattering arises only from the variation of n, i.e. $\sigma_{inc} \sim n$).

Similarly, the analysis of thermally excited surface waves in IGNATOVICH's paper /4.7/ yields a negligible probability of neutron heating, in contrast to a semi-quantitative treatment of this problem by FRANK /5.14/ which had indicated that Ray-leigh surface waves may be significant.

A common feature of such or other neutron interaction processes with thermal excitations subject to the Bose statistics is the temperature dependence which is determined by the occupation number $n_b = (e^{\hbar\omega/k_BT} - 1)^{-1}$. For very low frequency oscillations ($\hbar\omega \ll k_BT$), $n_b \sim T$, while for higher frequencies the temperature de-pendence is even stronger. The absence of any observed temperature dependence in the neutron containment experiments seems to be a strong argument against these hy-potheses.

On the other hand, temperature independence would be expected for the neutron interaction with low-energy Fermi systems, as in inelastic spin-flip scattering from nuclei with energetically favourable spin orientation. IGNATOVICH considers in /5.15/ the ortho-para-hydrogen transition in molecular hydrogen where both the scattering cross section and the energy splitting are very large. The calculation indicates that a hydrogen concentration of about 70 % in the surface would be necessary to account for the experimental values. Such a high concentration is quite unlikely.

The problem of acoustic and ultrasound vibrations was considered by GERASIMOV et al. /5.16/ and by IGNATOVICH /4.7/. The conclusion of /5.16/ is that realistic acoustic excitations lie several orders of magnitude below the level compatible with the experiments. The same result must be inferred from the result of /4.7/ since the ultrasound intensity which would produce the given oscillation amplitude of 1 Å at 10^{10} cps may be estimated to be fantastically high ($\approx$ 13 Bel!).

The effect of static surface roughness on the neutron absorption probability was investigated by IGNATOVICH /3.30, 2.7/, STEPANOV and SHELAGIN /5.10/ and STEPANOV /5.11/. The enhancement factor for macroscopic surface roughness follows from geo-metrical optics to be S/S_0, the ratio of the real surface S to the hypothetical plane surface S_0. In extreme cases S/S_0 may reach a value of 2. The case of small roughness amplitude $a \lesssim 50$ Å was analysed by perturbation theory /2.7/, while a model of a soft wall potential represented by a smoothed step function was introduced for narrow roughness features of larger amplitude /3.30/. The results of these analyses indicate that realistic roughness parameters cannot explain the large discrepancies, in particular for weak absorbers with predominant inelastic scattering since an enhancement of inelastic scattering should not alter its characteristic temperature dependence. Besides, the roughness of glass should be negligible as the agreement between experiment and elementary theory for the strong absorber boron glass demonstrates.

Similar objections apply also to the idea of absorption enhancement due to neutron tunneling into cavities within porous surfaces. IGNATOVICH considered in /5.17/ the probability of neutron resonance trapping in cavities of the right size. He came to the conclusion that for a random size distribution of spherical pores an enhancement factor of 10 would require a large volume porosity of 10 %.

The great unknown among all surface effects seems to be the role of surface contaminations, as the quantity and composition of adsorption layers on technical surfaces exposed to air, solvents, electrolytes or other agents is still largely the subject of speculations - and the object of surprise as some examples given by BIKERMAN (Ref. /5.18/, Ch. V) indicate. IGNATOVICH and STEPANOV have analysed in /5.12/ the effect of a homogeneous surface layer on the reflection loss. Both the screening or enhancement effect of the scattering potential of the layer on the neutron wave interaction with the substrate and the absorption in the layer itself were taken into account. The authors calculate that, e.g. a CCl_4 film of 14 Å thick would explain the experiments, by virtue of the exceptionally high capture cross section of chlorine. The amount of chlorine on technical surfaces is not known, although an analysis of ox   zes Al surfaces in /3.1/ by Ar-ion bombardment revealed only a negligible chlorine concentration. Contaminations with predominant inelastic scattering properties like hydrogen can hardly explain the experimental temperature independence of loss probabilities. Nevertheless, surface treatment has been shown to have some influence on the containment time /1.9/, at least up to a certain point. It would, however, seem strange that the achievable surface purity for a variety of different materials and over a wide range of temperature should happen to be such as to cause practically the same losses in each investigated case.

The "theoretical" curves in Fig.5.5 /5.6/ for the absorption coefficients $\bar{\mu}(v)$ in a copper trap are examples of attempts to apply the above theoretical considerations to an interpretation of the data. The curves were calculated under the following assumptions:

Curve 1: Elementary theory;

Curve 2: Elementary theory but with a 2.6 times enhanced capture cross section;

Curve 3: Roughness model of a soft potential step for the wall with an adjusted
width of 110 Å for the soft edge;

Curve 4: A hydrogenous contamination layer of 55 Å thick (at room temperature). Fig.5.5 shows that any one of curves 2, 3 or 4 could represent the data satisfactorily. The difficulties are much greater, however, with weakly absorbing substances where the discrepancies between the data and the elementary theory may exceed two orders of magnitude!

5.3.3 Search for Faults in the Elementary Theory

In the elementary theory of the index of refraction for slow neutrons (see Section 2.1) it is assumed that the effective field ψ_j acting on any scattering centre j at loaction $\vec{r}_j$ be the same as the total wave field $\Psi(\vec{r}_j)$ averaged locally, say over the range of a number of neighbouring atoms. However, it has been known since the works of FOLDY /2.1/, LAX /2.2, 3/ and EKSTEIN /2.4, 5/ that a rigorous multiple-scattering theory of the index of refraction should be based on the exact equations for $\Psi(\vec{r})$ and ψ_j

$$\Psi(\vec{r}) = \Phi(\vec{r}) - \sum_j \frac{e^{ik|\vec{r}-\vec{r}_j|}}{|\vec{r}-\vec{r}_j|} b_j \psi_j \tag{5.12}$$

and

$$\psi_j = \Phi(\vec{r}_j) - \sum_{\ell(\neq j)} \frac{e^{ik|\vec{r}_j-\vec{r}_\ell|}}{|\vec{r}_j-\vec{r}_\ell|} b_\ell \psi_\ell . \tag{5.13}$$

Eqs. (5.12) and (5.13) differ in the sense that $\psi(\vec{r})$ is determined by a superposition of the incident wave $\Phi(\vec{r})$ and the spherical wavelets originating from all atoms j (having scattering lengths b_j), whereas ψ_j is due to $\Phi(\vec{r}_j)$ and all scattering waves except for the wave originating from atom j.

LAX has postulated /2.3/ a strictly local relationship between the quantities $\bar{\Psi}$ and $\bar{\psi}$, averaged over a microscopic environment of a given point, i.e. $\bar{\psi} = C\bar{\Psi}$. As a consequence, the scattering potential should be modified by the same constant factor C. In the special case of a simple cubic crystal, EKSTEIN has derived for neutrons an equivalent of the Lorentz-Lorenz formula of optics, which shows that 1-C is real and very small of order $b/d \approx 10^{-4}$ where d is the lattice parameter (/2.5/, Eq. (33)). FRANK raised the question /5.19/ whether in the case of total reflection C might not have a finite imaginary part which would have the same effect on the reflection properties of an ultracold neutron as real absorption. Such a hypothesis, however, would be unable to account for the fate of those neutrons missing in the reflected beam.

IGNATOVICH and LUSHCHIKOV have analysed the case of total reflection from a perfect crystal in the framework of multiple scattering theory /2.7/. They came to the conclusion that no "anomalous diffusion through the interatomic space" should take place, and that, due to crystallinity, the absorption coefficient should increase only by a negligible fraction of order $\mu_0 k_{cr}^2 d^2 \approx 10^{-4} \mu_0$, where μ_0 is the absorption coefficient from the elementary theory. A similar theoretical investigation by STEPANOV /5.20, 21/ has also failed to reveal any significant fault in the simple theory of the optical potential for slow neutrons.

The persisting failure to account for the experimental containment losses has even stimulated the speculation that the anomaly observed may be an indication of a fundamental limitation of the maximum allowable size of an ultracold neutron wave-train to about 10^6 Å. Such a spatial limitation would entail an uncertainty in momentum which might allow the high energy fraction of the distribution to penetrate the wall with the desired probability. The generally accepted quantum theory, however, predicts as the only fundamental limit to the wave packet size ξ that due to the neutron's finite lifetime τ which is $\xi \approx \tau v = 5$ km for $v = 5$ m/s! IGNATOVICH proposed in /5.22/ a modification of the Schrödinger equation by an additional inhomogeneous term which would indeed permit one to construct nonspreading, localized wave packets with the required properties, resembling the dubious "singular solutions" of de BROGLIE (/5.23/, p. 100). Ignatovich postulates that the "packet size" vary inversely proportional to v in order to ensure a continuous transition from wave optics to geometrical optics. It should be pointed out that Ignatovich's hypothesis conflicts with such basic fundaments of quantum theory as the operator concept and the superposition principle, due to the inhomogeneity of the equation of motion. In the framework of quantum theory a "localization" (with respect to the relative coordinate) could only be conceived if the neutron were thought to be a composite system consisting of, say, two weakly bound particles. In addition, the postulated velocity dependence of the "neutron size" would contradict the basic symmetry law of nonrelativistic invariance of an observable length under a Galilean velocity transformation.

5.4 Containment of Faster Neutrons

In view of all the problems raised by the strange results of ultracold neutron storage experiments it seems gratifying that there exists another type of neutron containment which is obviously fully understood. In the experiments with the graphite bottle in Munich (/3.10/ and Fig.5.3) the curious observation was made that this bottle was not only capable of storing neutrons with velocities below the critical velocity for total reflection, but also faster neutrons. The reason for this is that graphite scatters very cold neutrons so strongly due to its inhomogeneous structure that it affects them in a similar way as some planetary atmospheres affect light: They reflect it diffusely with high efficiency. The reflectivity by virtue of the underlying diffusion process if commonly known as "diffuse reflectance" or "albedo".

The albedo is determined essentially by $\varepsilon = (\Sigma_{abs}/\Sigma_T)$, the ratio of the macroscopic absorption cross section Σ_{abs} due to nuclear capture and inelastic scattering to the total cross section $\Sigma_T = \Sigma_{abs} + \Sigma_{el}$ which includes the elastic scattering cross section Σ_{el} on inhomogeneities. For $\varepsilon \ll 1$ and isotropic illumination the albedo is approximately given by

$$A = 1 - \frac{4}{\sqrt{3}(1 - \overline{\cos\theta})}\sqrt{\varepsilon}, \tag{5.14}$$

where $\overline{\cos\theta}$ is the mean cosine of the scattering angle for elastic scattering. The albedo may reach considerable values for very inhomogeneous, weakly absorbing substances. For instance, an experimental value of 96 % was obtained at 87 K for $v = 13$ m/s, in good agreement with expectation. This opens the possibility of using graphite or even more inhomogeneous substances as efficient reflectors for very cold neutrons.

6. Possible Specific Applications of Very Low Energy Neutrons

Extensive research is presently under way in a number of laboratories in preparation of the two most promising experimental applications of very low energy neutrons: The search for a neutron electric dipole moment (EDM) using bottled neutrons and a precision measurement of the neutron lifetime for β-decay by neutron containment in a magnetic trap. Until now only the first results of cursory measurements by the Leningrad group on their EDM experiment became known /6.1/. Regarding the lifetime experiment planned by a group at Bonn University, only the principle of magnetic containment of neutral particles by the magnetic moment interaction has so far been demonstrated /6.2/. For this reason we shall discuss only briefly the physical ideas underlying these two projects, the principle experimental techniques and specifications, and the progress made in the experimental realization. Other, less-advanced proposals for possible future applications of very low energy neutrons, like a search for the neutron's electric charge, will be only touched upon.

6.1 Experimental Search for the Neutron Electric Dipole Moment

Since the first suggestion of PURCELL and RAMSEY in 1950 /6.3/ to search experimentally for electric dipole moments of elementary particles and nuclei as a means to check parity invariance, the case for experiments to detect a possible EDM of the neutron has been made repeatedly, for example more recently by SHAPIRO /1.5/ and by GOLUB and PENDLEBURY /6.4/.

   According to present general views the demonstration of a finite permanent EDM of the neutron - or any other elementary particle - would be of particular interest, not so much because of its implication of parity violation (which is now well established for the weak interaction), but rather because this would be the first direct evidence of time reversal noninvariance outside the system of K^0 mesons where CHRISTENSON et al. had first detected a CP and T violating decay mode in 1964 /6.5/. (It should be mentioned that the interpretation that the existence of an EDM would

violate T invariance applies if the EDM is thought to be due to a static electric charge distribution. RAMSEY has pointed out in 1958 /6.6/ that if magnetic
monopoles existed and the EDM was caused by rotating magnetic charges, the interpretation would have to be modified, because then the T symmetry would have to be
replaced by a combined TM symmetry, where M stands for magnetic charge conjugation.
Under these circumstances the existence of an EDM would be evidence for a violation
of TM invariance.)

The T violating decay of the long-lived K^0 meson has been confirmed and studied
in various respects during the last decade. A number of different theories have
been proposed as to the possible interaction responsible for the T violation. In
the framework of these theories various predictions were made relating to the expected order of magnitude of the neutron EDM. These estimates for the EDM, expressed
as a length D to be multiplied by the electron charge e, range from $D \cong 10^{-18} - 10^{-20}$
cm for the electromagnetic interaction as the assumed source of T noninvariance to
the prediction $D \cong 10^{-27} - 10^{-41}$ cm for Wolfenstein's "superweak" interaction theory.
Various weak interaction theories predict intermediary values.

Since the beginning of experimental search for a neutron EDM in 1950 by SMITH et
al. /6.7/ similar experiments have been performed many times (see, e.g. /6.4/). While
none has as yet succeeded in demonstrating a finite EDM the experimental upper limit
for a possible EDM has been significantly lowered by almost 5 orders of magnitude
in the course of the past 25 years, due to a gradual improvement in experimental
sensitivity. The most recent value obtained in 1973 - 75 at the Institute Laue-Langevin, Grenoble, by BAIRD, Dress, Miller, Pendlebury, Perrin and Ramsey is

$$D = (0.4 \pm 1.1) \times 10^{-24} \text{ cm} \tag{6.1}$$

(/6.8, 9/).

As the experimental limit was gradually lowered some theories which had predicted
larger EMD's were eliminated, and others were proposed instead. During the latest
experiment it seemed for a while that essentially only the genuine superweak interaction theory could survive the new experimental limit $D \leq 10^{-24}$ cm. However, the
recent estimates of LEE /6.10/ ($D \leq 10^{-23}$ cm), of FRENKEL and EBEL /6.11/ ($D \div 10^{-26}$
cm) and of MOHAPATRA and PATI /6.12/ ($D \leq 10^{-24}$ cm) obtained with certain unified
gauge models for the weak and electromagnetic interactions, are examples for the
continuing efforts of theoreticians to keep experimentalists busy in search for
further improvements in experimental techniques in order to perhaps reach the next
order of magnitude in sensitivity.

6.1.1 Principle and Limitations of EDM Experiments Using Magnetic Resonance

The most precise EDM experiments performed until now have used the split-coil magnetic resonance technique due to RAMSEY /6.13/ and pioneered by SMITH /6.7/. In

such an experiment slow neutrons in a beam are exposed to the action of static magnetic and electric fields, $\vec{B}$ and $\vec{E}$, with parallel or antiparallel mutual orientation. The measuring principle consists in trying to observe, by magnetic resonance, the slight change $\Delta\omega = \omega_+ - \omega_- = 4\mu_{el}E/\hbar$ in the Larmor precession frequency

$$\omega_\pm = \frac{2}{\hbar} (\mu_{mag}B \pm \mu_{el}E),\tag{6.2}$$

which is expected to occur as a result of reversing the direction of the strong $\vec{E}$ field relative to the weak $\vec{B}$ field, if the neutron has a finite electric dipole moment $\vec{\mu}_{el} = e\vec{D}$ in addition to its magnetic dipole moment $\vec{\mu}_{mag}$. (In these considerations it is assumed that any observable, effective permanent dipole moment should lie in the direction of the spin since the spin is the only quantity characterizing the particle "orientation". Classically speaking, any perpendicular component would average out to zero by the particle's rotation.)

As shown in a comprehensive way by GOLUB and PENDLEBURY /6.4/, there are two essential factors limiting the sensitivity of this type of EDM experiment (and in a similar way also of any other type): One is connected with the uncertainty principle limitation and the other with the so-called $(\vec{v} \times \vec{E})$ effect.

The uncertainty principle $\Gamma t = 1/2$ relates the minimum line width Γ of magnetic resonance with the time t spent by the particle in the region of the homogeneous $\vec{B}$ and $\vec{E}$ fields where the resonance condition is fulfilled. The extent to which the resonance curve may be resolved is determined by statistics. This leads to the limit

$$D > \frac{\hbar}{2eEt\sqrt{N}}\tag{6.3}$$

for a detectable electric dipole length. N is the total number of "useful" neutrons counted in the experiment. (The "usefulness" depends largely on the efficiency of production and analysis of neutron beam polarization.)

The $(\vec{v} \times \vec{E})$ problem refers to a spurious effect of the $\vec{E}$ field, due to the interaction of the magnetic dipole moment with the magnetic field $\vec{B}_v = -(\vec{v} \times \vec{E})/c$ which the neutron "sees" in its moving frame as it travels with velocity $\vec{v}$ through the applied electric field. This interaction vanishes (in first order) for exact alignment of the $\vec{B}$ and $\vec{E}$ fields. Furthermore, it may in principle be separated from a genuine EDM interaction by making use of the change of sign of $(\vec{v} \times \vec{E})$ under a reversal of velocity. Nevertheless, it is impossible in practice to eliminate the $(\vec{v} \times \vec{E})$ effect entirely due to the unavoidable geometrical uncertainties of the experimental apparatus. Indeed, the error for D given in (6.1) seems to be determined largely by the uncertainty of the $(\vec{v} \times \vec{E})$ correction.

6.1.2 Advantage and Feasibility of the Use of Ultracold Neutrons

The major attraction of the use of stored ultracold neutrons, instead of a beam of slow neutrons, in a magnetic-resonance-type EDM experiment is the possibility to

reduce significantly both the uncertainty principle limitation and the ($\vec{v}$ x $\vec{E}$) problem. The reason is that, on the one hand, the neutron interaction time t with the
fields may be enhanced by about three orders of magnitude to, say, 30 s as compared
to about 0.02 s for the recent experiment at Grenoble where a beam of neutrons of
$\simeq$ 100 m/s was used and the apparatus was 2 m long /6.8/. On the other hand, the
effective neutron velocity determining the ($\vec{v}$ x $\vec{E}$) effect is reduced by as much as
four orders of magnitude, mainly because the mean velocity vector averaged over the
random zig-zag path traced by a neutron during containment, becomes very small.
Therefore, the requirements imposed on field parallelity may be considerably re-
laxed. Similarly, a trapped neutron averages very efficiently over spatial field
inhomogeneities which are thus much less critical than in a beam-type experiment.

These qualitative considerations in favour of a high-sensitivity EDM experiment
using stored ultracold neutrons have been noted first by SHAPIRO in 1968 /1.5/. They
were later borne out in detail in a number of specific feasibility studies and preparatory experiments by the various groups planning to do such experiments in Leningrad /6.1, 14/, in Dubna /6.15 - 21, 6.22, 23/ and in Grenoble /6.4, 8, 24/.

Among these projects that of the group of Professor Lobashov in Leningrad has
until now been advanced farthest. Fig.6.1 shows the schematic features of their
apparatus /6.14/. Ultracold neutrons travel up the mirror neutron pipe 1 which was
described in Section 3.4.1 (Fig.3.5), and are polarized on passing through a magnetized iron foil 2. (The efficiency of such a polarizer was determined experimentally
as described below.) The polarized neutrons enter the magnetic resonance spectrometer which is located in the centre of a three-layer magnetic shield 3. The resonance storage chamber consists of a BeO-coated quartz cylinder 4 of 50 cm in diameter
with flat Be ends 5 serving as electrodes which produce an E field of 20 - 25 kV/cm
across the chamber height of 6 cm. A highly uniform $\vec{B}$ field of 0.1 G parallel to $\vec{E}$
is generated by optimally arranged Helmholtz coils 6. Using the Ramsey method of
separate oscillating coils, the polarized neutrons are passed through the first
coil 7 which generates an oscillating magnetic field directed perpendicular to $\vec{B}$
and set to the neutron Larmor frequency ν_0 in the applied field $\vec{B}$. Due to the action
of the oscillating field whose strength B_1 is properly adjusted to the neutron tran-
sit time through the coil, the neutron spin is rotated to a direction perpendicular
to the $\vec{B}$ and $\vec{E}$ fields. During the dwell time of 2 to 3 s in the chamber, the neutron precesses about the stationary fields until it happens to hit the exit- - or
entrance- - window. On passing through the second oscillating field coil 8 which is
identical to the first one and generates the same field B_1 but with a phase difference of 90^0, a small phase angle accumulated by the neutron spin in the resonance
chamber, relative to the phase of the oscillating field which serves as a clock,
is very sensitively converted to neutron polarization. The phase difference of 90^0
between the two oscillating fields ensures that the working point lies on the steep

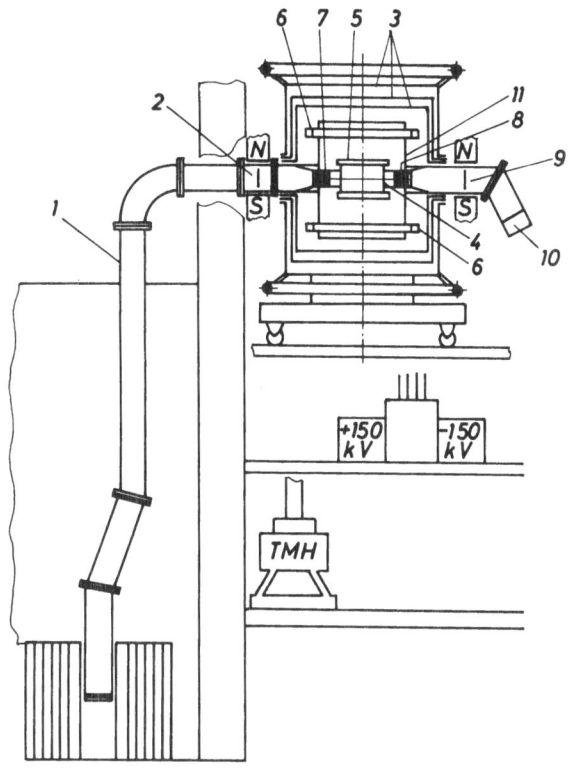

Fig.6.1. Scheme of the apparatus designed at the reactor VVR-M, Leningrad, for an experimental search for the neutron's electric dipole moment /6.14/. *1* vertical ultracold neutron guide tube; *2* polarizing iron foil; *3* three-layer magnetic shield; *4* BeO-coated glass cylinder; *5* beryllium lids serving as high-voltage electrodes; *6* Helmholtz coils; *7* first oscillating field coil; *8* second oscillating field coil; *9* iron foil for polarization analysis; *10* ^{3}He detector; *11* vacuum chamber

edge of the resonance curve (Fig.6.3). The polarization of departing neutrons is analysed by transmission through the magnetized iron foil *9*. Neutron detection is accomplished by a ^{3}He detector *10* (see Section 3.5). The direction of the $\vec{E}$ field may be reversed every 100 s.

Similar apparatus are in preparation for the planned EDM experiments in Dubna and Grenoble, except for a possible variation in the mode of operation. While in the "passage variant" used in the Leningrad design (Fig.6.1), the resonance chamber is an open cavity through which the neutrons pass in a random way, in the "storage variant" the neutrons are admitted to the resonance cell through a valve. This valve will be closed for the desired storage duration, say, of about 30 s, and then re-opened thus allowing the surviving neutrons to discharge through the same port where they had entered /6.16/. A design allowing both modes of operation is also possible /6.24/. A modification of the oscillating field arrangement has been pro-posed for the Dubna project "Tristom" where the enclosed neutrons are to be sub-jected to oscillating field pulses of appropriate duration after admission to the storage cell and before departure /6.22/.

The requirements imposed by the criterion of a given experimental sensitivity to an EDM, on parameters like the oscillator frequency control, parallelity of the $\vec{E}$ and $\vec{B}$ fields, and the spatial uniformity, time stability and $\vec{E}$-field independence of the $\vec{B}$ field, have been discussed repeatedly /6,15, 16, 18, 20; 6.8; 6.23; 6.24; 6.14/. According to SMITH et al. /6.24/ it should be possible to meet the conditions for a sensitivity of $\Delta D \approx 10^{-26}$ cm for the Grenoble experiment with reasonable pre-cautions, and with much higher efforts a sensitivity of $\approx 10^{-28}$ cm might ultimately be reached.

(The high level of such a sensitivity may be appreciated by the following con-sideration: If one imagined the EDM to be produced by two elementary charges $\pm$e separated by the distance D then a sensitivity of 10^{-26} cm would correspond to a separation amounting to only $\approx 10^{-13}$ of the neutron size, that is as little as 1 μm in relation to the earth's diameter!)

6.1.3 Preparatory Experiments and First Results

A number of auxiliary experiments have so far been performed in preparation for EDM experiments using ultracold neutrons.

6.1.3.1 Polarization and Spin Flip of Ultracold Neutrons

EGOROV et al. reported in /3.9/ on their investigations of the use of ferromagnetic foils for polarizing ultracold neutrons. As discussed in Sections 2.1 (2.8) and 4.3 the neutron scattering potential of a magnetized ferromagnet,

$$U_\pm = \frac{2\pi\hbar^2}{m} \ Nb \pm \mu B,$$

depends on the neutron spin orientation parallel (+) or antiparallel (-) to mag-netization. Therefore, neutrons passing at perpendicular incidence through a mag-netized ferromagnetic foil will be polarized antiparallel to magnetization, if their energy lies in the interval between U_+ and U_- (or 0 if $U_- < 0$). Optimum conditions are achieved for $U_- \cong 0$, since in this case no refraction, and hence no partial reflection, of the transmitted spin state occurs. Suitable materials in this re-spect are, for example, the alloy 54 % Fe with 46 % Co or the isotope mixture 20 % ^{56}Fe with 80 % ^{54}Fe.

The Leningrad group performed an experiment /3.9/ to determine the polarization
efficiency of various substances by the arrangement sketched in Fig.6.2. Neutrons
are passed successively through identical polarizer and analyser foils 1 and 2. In
the region between polarizer and analyser the neutron spin may be flipped relative
to the magnetic field. Then the ratio of the count rates with and without spin flip
is a measure for the polarization efficiency.

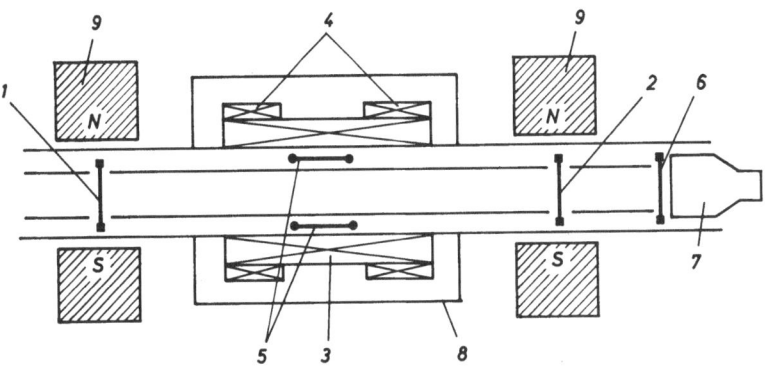

Fig.6.2. Arrangement of apparatus for determining the polarization efficiency for
ultracold neutrons /3.9/. 1 polarizer foil; 2 analyser foil; 3 selonoid producing
a constant magnetic field; 4 coils for generation of a magnetic field gradient;
5 oscillating field coils; 6 screen for determining the background; 7 detector;
8 magnetic screen; 9 magnets for magnetization of polarizer and analyser

    Since very slow neutrons follow adiabatically all reasonable spatial changes
in the direction of applied magnetic fields, conventional methods of neutron spin
flip (or depolarization) cannot be used for ultracold neutrons. Instead Egorov et
al. applied the method of "fast adiabatic spin flip" which is well known in nuclear
magnetic resonance. The neutrons move through a static magnetic field (generated
by the coils 3 and 4 in Fig.6.2) with a field gradient (say, negative) along the neutron path. Perpendicular to this field an oscillating field is supplied by the coil
5, with a frequency set to the neutron Larmor frequency in the static field H_0 at
the centre of the oscillating field region. In the reference frame rotating with
the oscillating field (which may be thought to be composed of two counter-rotating
fields) the effective field, which includes a field term $-\vec{H}_0$ arising from the transformation to the rotating system (as RABI et al. have first noted in /6.25/), gradually reverses its orientation from a direction along the static field on entrance

to the oscillating field region, to the opposite direction on exit. (For a positive field gradient the reversal proceeds in the opposite sense.) A neutron traversing this region sufficiently slowly will follow the field reversal adiabatically, and thus its spin will be flipped. Explicitly, the condition of adiabaticity reads

$$v\left|\frac{dH}{dx}\right| \ll 2\pi\gamma H_{var}^2 \qquad\qquad (6.4)$$

where dH/dx is the field gradient, H_{var} is the amplitude of the rotating field (= 1/2 of the oscillating field) and $\gamma = 2\mu/h$. Quantitative calculations of the spin-flip efficiency ε were reported by TARAN /6.26/.

Assuming $\varepsilon = 1$ Egorov et al. determined a polarization efficiency of $\approx$ 75 % for iron foils, nearly independent of the isotope composition and preparation.

TAYLOR has recently measured the velocity dependence of polarization by FeCo films at the neutron turbine in Munich /6.27/. In this experiment the spin-flip efficiency was determined separately by arranging two spin flippers in series (according to a proposal by TARAN /6.28/). Values of $\varepsilon > 0.9$ were observed in the velocity region of interest. In addition, the data evaluation was modified compared with the Leningrad work also in order to take account of the difference in the processes giving rise to incomplete polarization, between the transmission and reflection geometry. In reflection, imperfect polarization is usually assumed to be due to partial reflection of the wrong spin state, while in transmission the depolarizing effects of magnetization fluctuations within the foil seem to be dominant. In spite of these details, the polarization efficiencies obtained by Taylor are consistent with the Leningrad data.

6.1.3.2 Other Preparatory Work

Because of the need of very uniform and stable magnetic fields, much effort was devoted to the calculation and testing of efficient magnetic shields /6.18, 20; 24/. In addition, the optimum designs for homogeneous field generation were studied /6.14, 18/. Other important tests were concerned with the possible depolarization of neutrons in the total reflection process. No depolarization was observed even for a ferromagnetic Ni reflector /6.14/. The leak currents in BeO were found to be tolerable in view of its use as insulator wall in the storage cell of the resonance spectrometer /6.24/.

6.1.3.3 First Results

As a first result of experiments with the EDM apparatus in Leningrad an ultranarrow line width Γ = 0.2 - 0.3 Hz of the resonance curve shown in Fig.6.3 was observed /6.14/, in good agreement with the calculated dwell time of the neutrons in the chamber. This value is to be compared with the $\approx$ 100 times broader resonance line in the recent beam experiment in Grenoble /6.8/.

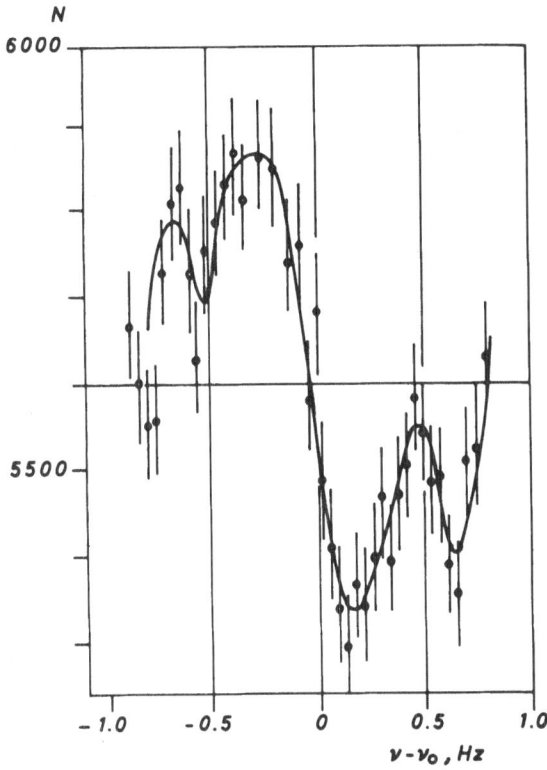

Fig.6.3. Magnetic resonance curve obtained in the EDM spectrometer in Leningrad with BeO-coated quartz glass /6.14/. The ultracold neutron counting rate is plotted versus the oscillating field frequency. Resonance frequency ν_0 = 312.5 Hz; phase difference between the two oscillating field coils: δ = 90°

A first cursory EDM experiment during a measuring period of one week at Leningrad yielded the result

$$D = (0.8 \pm 2.6) \times 10^{-23} \text{ cm}$$

/6.1/ which indicates the high potentialities of the method if all its odds are fully utilized.

6.2 Magnetic Neutron Containment and a Possible Lifetime Experiment

The anomalously high reflection losses of neutrons in material bottles seem to rule out, for the time being, the precision measurement of the neutron lifetime for β-decay envisaged by ZEL'DOVICH in 1959 /1.3/ (except, perhaps, if "hypercold neutrons" in a flat bowl-like trap are used, as proposed by KOSVINTSEV et al. /6.29/. For such a geometry the theoretical containment time is maximal for neutrons hopping over the ground in little bounces). It is hoped, however, that a lifetime experiment may be feasible using neutrons trapped by magnetic fields.

It is well known from atomic and molecular beam research that neutral particles with a magnetic moment μ may be guided and focused by suitable magnetic fields $\vec{B}(\vec{r})$, due to the interaction energy $V = -\vec{\mu}\cdot\vec{B}(\vec{r})$. Under the condition of adiabaticity, i.e., if all orientational changes of $\vec{B}(\vec{r})$ as "seen" by the particle moving through the magnetic field occur very slowly compared with the Larmor frequency, the magnetic quantum number which determines the spin orientation of the particle with respect to the magnetic field remains unchanged. Then the force acting on, say, a neutron moving in an inhomogeneous magnetic field is given by

$$\vec{F} = - \nabla V = \pm \mu \nabla B(\vec{r}) \tag{6.5}$$

where the + (-) sign holds for neutron spin antiparallel (parallel) to the field. (The gyromagnetic ratio of the neutron is negative!)

PAUL has proposed in 1951 a magnetic lens for neutrons on the basis of a magnetic sextupole field generated by three pairs of magnetic poles distributed symmetrically about an axis /1.7/. For such a configuration, $B \sim r^2$ for not too large distances r from the axis. Thus, the particle is subjected to a two-dimensional harmonic potential which acts as a focusing lens for the spin state parallel to $\vec{B}$ and as a dispersing lens for the other spin state.

6.2.1 Principle and Feasibility of Magnetic Neutron Containment

VLADIMIRSKY proposed in 1960 to use the repulsion of neutrons with spin aligned with a magnetic field, from regions of strong field for neutron confinement /1.8/. Both the confinement in two dimensions along a focusing channel and in three dimensions in closed systems were considered. Among the various magnetic field configurations shown to be suitable for constructing a magnetic neutron container, the focusing channel bent to a toroidal shape, as sketched in Fig.6.4, offers the advantage that it can hold a larger section in neutron phase space, and hence a larger number of neutrons per unit volume, than single-connected cavities enclosed by magnetic mirrors, because the momentum limitation along the channel axis is considerably wider than in the direction normal to a magnetic wall.

The cylinder symmetric magnetic (2n)-pole field required for a straight neutron channel may be generated by 2n linear parallel conductors with currents I of alternately opposite direction, arranged along the axis z in a symmetric way at a radial distance r_0. The magnetic field for such a configuration lies in the (x,y)-plane perpendicular to z. For $r \leq 0.5\ r_0$ a good approximation for the radial and azimuthal field components is

$$B_r = B_0 \ (r/r_0)^{n-1} \sin n\phi$$
$$B_\phi = B_0 \ (r/r_0)^{n-1} \cos n\phi \tag{6.6}$$

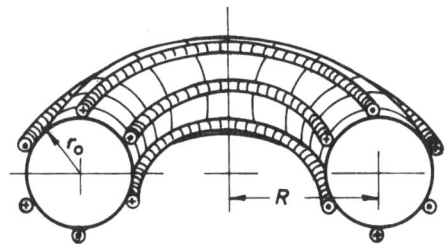

Fig.6.4. Cross section through a sextupole torus formed by six circular conductors /6.2/

where $B_0 = 4nI/r_0$, and ϕ is the azimuthal angle referred to the angular position of a conductor. Hence

$$B(r) = B_0 (r/r_0)^{n-1}. \tag{6.7}$$

Focusing may be achieved along the z-axis for any number of pole pairs $n \geq 3$, where $n = 3$ corresponds to the sextupole lens.

As the height of the potential wall is μB_{max}, the energy of motion in the radial direction must be

$$E_r = \frac{m}{2} v_r^2 < \mu B_{max}, \tag{6.8}$$

where B_{max} is the maximum usable field. For neutrons, $E_r/B_{max} < 6.03 \times 10^{-8}$ eV/T.

The situation when the axis is bent to a circle and the field configuration to a torus with large radius R, as in Fig.6.4, has been considered in detail by MARTIN /6.2/. The neutron remains in the torus under the condition that the centrifugal force is compensated. For a sextupole field, this condition limits the axial velocity to values less than

$$v_{z,max} = v_{r,max} \sqrt{1 + R/r_0} \tag{6.9}$$

which may be considerably larger than the maximum allowable lateral velocity $v_{r,max}$. This is the basis for the desired gain in the volume of neutron phase space suitable for storage.

Since only neutrons of one spin state may be stored in a magnetic trap the problem of spin flipping due to a breakdown of adiabaticity is of great significance. The adiabatic condition may be violated when a neutron crosses a region where the magnetic field is very small and changes its orientation rapidly. This situation has been considered by VLADIMIRSKY /1.8/. He obtained for the probability of spin flip

$$w = e^{-\pi\omega_L t/2} \tag{6.10}$$

where ω_L is the Larmor frequency in the minimum field on the neutron trajectory and
t is the effective time for reorientation of the field. The requirement that w be
sufficiently small for a lifetime experiment imposes a restriction on the minimum
allowable field strength in a magnetic neutron trap, which can be met with reasonable precautions. MATORA pointed out that, although in a multipole field B = 0 on
the axis, for any odd number of pole pairs (as in the sextupole) the field main-
tains its direction along any straight trajectory through the axis so that spin flip
is unlikely to occur even in this region /6.30/. IGNATOVICH considered in /6.31/
the influence of gravity on the neutron motion in a sextupole field. Gravity (or
other similar forces like the centrifugal force) give rise to mixing of the spin-
stable states with large angular momentum about the z-axis, which correspond to
neutron motion in the high fields far from the axis, with states with low angular
momentum which decay more quickly. The results indicate that this effect may be
significant if long storage times are required in experiments to determine the neutron lifetime for β-decay.

Such an experiment is presently planned at the high-flux reactor in Grenoble.
A group under Professor Paul at Bonn University has built a superconducting storage
ring as shown in Fig.6.5 /6.32/. The storage ring of about 1 m in diameter has been
simplified by removing the two inner coils of the sextupole which are rendered un-
necessary by the centrifugal barrier. The outer two of the necessary four conductors
are split into two coils each with the same current directions. This arrangement
provides a nonharmonic decapole component ($\sim r^4$) in the field which helps to control
possible resonances due to the coupling of the transverse to the longitudinal motion
in an imperfect torus, by limiting the amplitude of the transverse neutron "betatron"
oscillations. The problem of depolarization by spin flip is largely avoided since
the field-free region may be excluded from the working volume which is displaced to
larger radii by the centrifugal force. The coils are designed to generate a maximum
field of 3.5 T and a maximum gradient of 1.2 T/cm. This should enable the ring to
store neutrons in an azimuthal velocity band from 10 to 20 m/s and in a region of
lateral velocities ≤ 4 m/s. Neutrons are to be injected into the ring by a strongly
curved section of guide tube which may be withdrawn from the working volume after
filling. The preparatory work of KÜGLER /6.33/ indicates that plastic scintillation
counters may be used to detect the decay electrons.

A variety of different field configurations for other kinds of possible magnetic
traps suitable for even slower neutrons have been discussed. Among these are the
following:

a) The three-dimensional harmonic oscillator potential generated by the coil
configuration of Fig.6.6 /6.2/;

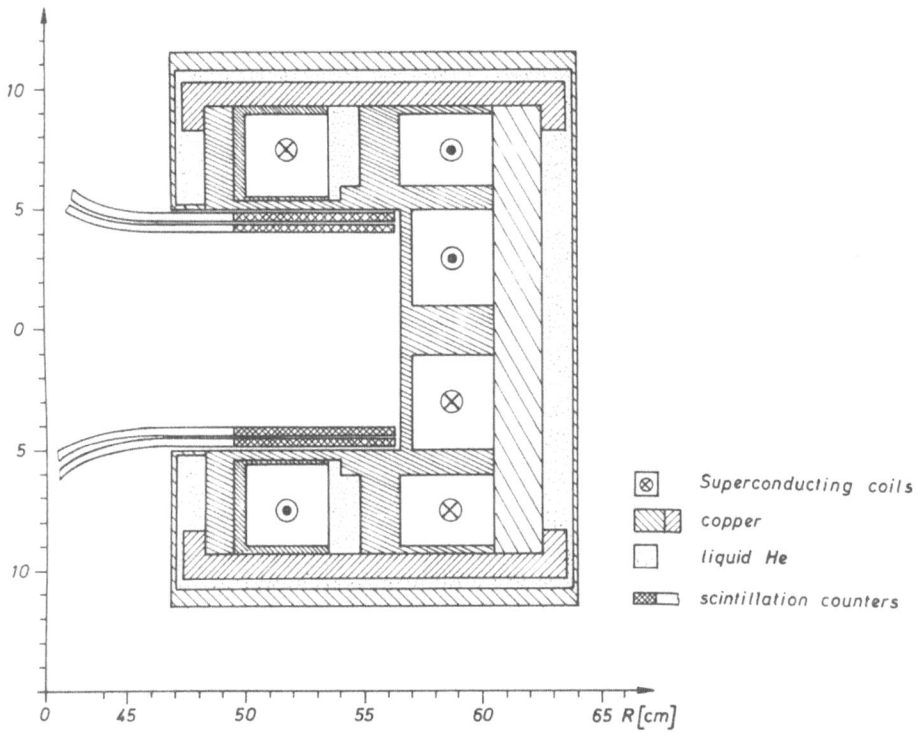

Fig.6.5. Coil configuration for the neutron magnetic storage ring built by Professor Paul's group at Bonn University /6.32/

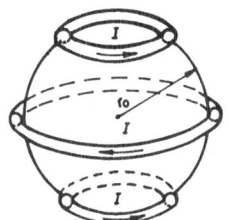

Fig.6.6. Spherical magnetic bottle generated by three circular coils arranged in such a way that the cross section forms a hexagon /6.2/

b) The field of a conductor which theoretically allows neutron trapping in helical trajectories about the conductor axis, due to the attractive interaction of strong fields with the neutron spin state antiparallel to the field /1.8/;

c) A magnetic "bowl" with magnetic mirrors defining the bottom and the cylindrical side wall, in which the neutrons are to be confined by gravity /6.34/.

6.2.2 First Experiments and Results

MARTIN has constructued and successfully operated /6.2/ at Bonn University a sextupole sphere generated by the arrangement of superconducting coils shown in Fig.6.6. The test containment experiments performed with sodium atoms demonstrated that the magnetic confinement of neutral particles is possible.

KOSVINTSEV et al. have studied in Dimitrovgrad both the transmission of ultracold neutrons through the magnetic barrier established by a solenoid /2.14/ and the containment lifetime in a material bottle closed on one end by a magnetic field of the same type /2.13/. The experiments showed that the magnetic barrier is of the expected magnitude and its sign depends on the spin orientation. In addition, it was demonstrated that depolarization may be efficiently suppressed by the application of a guide field.

6.2.3 The Case of an Improved Neutron Lifetime Experiment Using Stored Neutrons

The decay of the free neutron into a proton, electron, and antineutrino is the only known β-decay process where the (Fτ)-value can be interpreted with regard to the weak-interaction constant G_A characterizing transitions of the axial-vector, Gamov-Teller type (see, for instance, the review on the beta decay of the neutron by ERO-ZOLIMSKY /6.35/). (τ is the mean lifetime and F is the generally accepted function of the beta spectrum end-point energy introduced by Fermi.) Since the corresponding constant G_V for the vector-type Fermi interaction is known with high precision from the (Fτ)-values of other nuclei, the study of the neutron decay allows one to determine the fundamental constant $|\lambda| = G_A/G_V$ of the pure V-A weak-interaction theory which seems to be consistent with all available data on angular correlations. For this reason an accurate measurement of the neutron lifetime is of high interest.

The existing experimental data on the neutron lifetime were all obtained by the method of counting the decay products emitted by a thermal neutron beam. This technique requires two independent absolute measurements, namely, of the number of decays, which are counted by electron or proton detectors, and of the number of neutrons in the working volume of the beam, which are counted by neutron detectors. Therefore, one has to deal with uncertainties with regard both to the geometry and the absolute detector efficiencies. In view of these problems it seems to be not too surprising that the two experiments contending for the highest precision, that of SOSNOVSKY et al. of 1959 /6.36/ and that of CHRISTENSEN et al. of 1967 /6.37/, differ by as much as 10 % which lies considerably outside the quoted errors of respectively 3 % and 1.5 %.

A lifetime experiment using neutrons in a trap rather than in a beam, on the other hand, would essentially consist of a simple relative measurement of the decay rate as a function of time, as with an ordinary radioactive substance. Thus we may expect an improvement in the experimental value for the neutron lifetime, provided that the theoretical low containment losses in magnetic traps can be achieved in practice.

6.3 Possibility of an Experiment in Search for the Neutron's Electric Charge

Compared to the proposed EDM and lifetime experiments with stored neutrons, only little attention has so far been paid to the possible application of ultracold neutrons in a high-sensitivity search for the electric charge of the neutron. It is well known that a finite charge of the neutron should exist if the leptonic and the baryonic unit charge are not exactly equal.

KASHUKEEV has pointed out /1.6/ that the sensitivity of a direct experiment to observe a possible neutron deflection by a strong electric field should be significantly enhanced by the application of ultracold neutrons which may be kept in the region of the electric field for a long time compared to the traversal time of thermal neutrons used in previous direct experiments (the most accurate of which was performed by SHULL et al. in 1967 /6.38/). This qualitative judgement seems to be supported by a recent design and feasibility study for such a project by KASHUKEEV and YANEVA /6.39/. However, it should be emphasized that the indirect evidence from the observed neutrality of atoms and celestial bodies has set much lower margins to a possible neutron charge than any known direct method can apparently provide.

Acknowledgement

In conclusion I would like to thank all the colleagues who were so kind as to inform me about the course and the results of their investigations.

References

1.1 A. Steyerl: Phys. Letters 29B, 33 (1969)
1.2 E. Fermi: Nuclear Physics (Univ. of Chicago Press 1950)
1.3 Ya.B. Zel'dovich: Soviet Phys. JETP 9, 1389 (1959)
1.4 V.I. Lushchikov, Yu.N. Pokotilovsky, A.V. Strelkov, F.L. Shapiro:
     Soviet Phys. JETP Letters 9, 23 (1969)
1.5 F.L. Shapiro: Soviet Phys. Uspekhi 11, 345 (1968)
1.6 N.T. Kashukeev: Dokl. Bolgar. Akad. Nauk 24, 717 (1971)
1.7 W. Paul: In Proc. Int. Conf. on Nuclear Physics and Physics of Fundamental
     Particles, Chicago, ed. by J. Orear, A.H. Rosenfeld, R.A. Schluter (Univ.
     of Chicago Press 1951) p. 172
1.8 V.V. Vladimirsky: Soviet Phys. JETP 12, 740 (1961)
1.9 F.L. Shapiro: Conf. on Nuclear Structure Study with Neutrons, Budapest (1972);
     In Proc. Nuclear Structure Study with Neutrons, ed. by J. Erö and J. Szücs
     (Plenum Press New York/London 1975) p. 259; JINR Preprint R3-7125 (Dubna 1973)

1.10 I.I. Gurevich, L.V. Tarasov: Fizika neitronov nizkikh energii (Nauka, Moskva 1965); Engl. transl. Low-Energy Neutron Physics (North-Holland, Amsterdam 1968) p. 274

1.11 I.M. Frank: JINR Preprint R3-7809 (Dubna 1974)

2.1 L.L. Foldy: Phys. Rev. 67, 107 (1945)

2.2 M. Lax: Rev. Mod. Phys. 23, 287 (1951)

2.3 M. Lax: Phys. Rev. 85, 621 (1952)

2.4 H. Ekstein: Phys. Rev. 83, 721 (1951)

2.5 H. Ekstein: Phys. Rev. 89, 490 (1953)

2.6 V.K. Ignatovich: JINR Preprint R4-6553 (Dubna 1972)

2.7 V.K. Ignatovich, V.I. Lushchikov: JINR Communication R3-8795 (Dubna 1975)

2.8 R.Lenk: phys. stat. sol. (b) 69, 369 (1975)

2.9 E. Fermi, W.H. Zinn: Phys. Rev. 70, 103 (1946)

2.10 H. Maier-Leibnitz, T. Springer: J. Nucl. Engergy A/B 17, 217 (1963)

2.11 J. Christ, T. Springer: Nukleonik 4, 23 (1962)

2.12 L.V. Groshev, V.N. Dvoretsky, A.M. Demidov, Yu.N. Panin, V.I. Lushchikov, Yu.N. Pokotilovsky, A.V. Strelkov, F.L. Shapiro: Phys. Letters 34B, 293 (1971)

2.13 Yu.Yu. Kosvintsev, Yu.A.Kushnir, V.I.Morozov, A.V. Strelkov: Ultracold Neutron Containment in a Vessel with a Magnetic Wall, 3rd Conf. on Neutr. Phys. (Kiev 1975)

2.14 Yu.Yu. Kosvintsev, Yu.A. Kushnir, V.I. Morozov: Zh. Techn. Fiz. Pis'ma 2, 293 (1976)

2.15 A. Steyerl: 2nd International School on Neutron Physics (Alushta); In Proceedings JINR D3-7991 (Dubna 1974) p. 42

2.16 A. Steyerl: 38. Physikertagung (Nürnberg); In Proceedings "Plenarvorträge 11, Physik 1974" (Deutsche Physikalische Gesellschaft 1974) p. 86

2.17 R. Lermer, A. Steyerl: phys. stat. sol (a) 33, 531 (1976)

2.18 W. Schmatz, T. Springer, J. Schelten, K. Ibel: J. Appl. Cryst. 7, 96 (1974)

2.19 R. Golub, P. Carter: Nucl. Instr. and Methods 91, 205 (1971)

2.20 A. Steyerl: "Vorläufiger Bericht über zwei Verfahren zur Gewinnung von subthermischen Neutronen im Energiebereich $10^{-5} - 10^{-7}$ eV"; FRM-Report PDTUM-FRM-113 (Technical University, Munich 1966)

2.21 A. Steyerl: Nucl. Instr. and Methods 125, 461 (1975)

2.22 N.T.Kashukeev: Dokl. Bolgar. Akad. Nauk 23, 1473 (1970)

2.23 N.T.Kashukeev, G.A. Stanev, N.B. Yaneva, D.S. Mircheva: Nucl. Instr. and Methods 126, 43 (1975)

2.24 A. Steyerl, H. Vonach: Z. Physik 250, 166 (1972)

2.25 W. Dilg, W. Mannhart, E. Steichele, P. Arnold: Z. Physik 264, 427 (1973)

2.26 W. Dilg, W. Mannhart: Z. Physik 266, 157 (1974)

2.27 E.Z. Akhmetov, V.V. Golikov, D.K. Kaipov, V.A. Konks, A.V. Strelkov: JINR Communication R3-8470 (Dubna 1974)

2.28 H. Maier-Leibnitz: Nukleonik 8, 61 (1966)

2.29 V.A. Namiot: Dokl. Akad. Nauk USSR 221, 76 (1973) (Sov. Phys. Dokl. 18, 481 (1974))

2.30 R. Golub, J.M. Pendlebury: Phys. Letters 53A, 133 (1975)

3.1 A. Steyerl: Nucl. Instr. and Methods 101, 295 (1972)

3.2 R. Golub: Phys. Letters 38A, 177 (1972)

3.3 V.V. Golikov, V.I. Lushchikov, F.L. Shapiro: Zh. Eksp. i Teor. Fiz. 64, 73 (1973) (Sov. Phys. - JETP 37, 41 (1973))

3.4 G.D. Porsev, A.P. Serebrov: LINP Preprint 38 (Leningrad Institute of Nuclear Physics 1973)

3.5 J.M. Robson, D. Winfield: Phys. Lett. 40B, 537 (1972)

3.6 J.M. Robson: Canad. J. Phys. 54, 1277 (1976)

3.7 J.C. Bates, S. Roy: Nucl. Instr. and Methods 120, 369 (1974)

3.8 Yu.Yu. Kosvintsev, E.N. Kulagin, Yu.A. Kushnir, V.I. Morozov, A.V. Strelkov: Extraction of UCN from the High-Flux Reactor SM-2; Submitted to Nucl. Instr. and Methods (1976)

3.9 A.I. Egorov, V.M. Lobashov, V.A. Nazarenko, G.D. Porsev, A.P. Serebrov: Sov. Phys. J. Nucl. Phys. 19, 147 (1974) (Yad. Fiz. 19, 300 (1974)); V.M. Lobashov, G.D. Porsev, A.P. Serebrov: LINP Preprint No. 37 (Leningrad 1973)

3.10 A. Steyerl, W.-D. Trüstedt: Z. Phys. 267, 379 (1974)
3.11 S. Chandrasekhar: Radiative Transfer (Dover Publ. New York 1960) p. 125
3.12 V.V. Sobolev: A Treatise on Radiative Transfer (Van Nostrand, Toronto/New York/London 1963) p. 147
3.13 L.V. Groshev, V.N. Dvoretsky, A.M. Demidov, S.A. Nikolayev, Yu.N. Panin, V.I. Lushchikov, Yu.N. Pokotilovsky, A.V. Strelkov, F.L. Shapiro: JINR Preprint R3-7282 (Dubna 1973)
3.14 E.Z. Akhmetov, D.K. Kaipov, V.A. Konks, V.I. Lushchikov, Yu.N. Pokotilovsky, A.V. Strelkov, F.L. Shapiro: JINR Preprint R3-7457 (Dubna 1973)
3.15 P. Ageron, G. Germain: Private communication (1976)
3.16 D. Marx, B. Alefeld, K. Berndorfer, A. Steyerl: Bericht über neuere Entwicklungen zur Experimentiertechnik mit langsamen Neutronen in München; in Diskussionstagung über Neutronenphysik an Forschungsreaktoren (KFA Jülich, 25-28 April, 1967)
3.17 E. Tunkelo, A. Palmgren: Nucl. Instr. and Methods 46, 70 (1967)
3.18 A. Palmgren: Acta Polytechn. Scand. Ph 52 (1968)
3.19 A.V. Antonov, D.E. Vul', M.V. Kazarnovsky: JETP Lett. 9, 180 (1969)
3.20 N.T. Kashukeev, D.S. Mircheva, N.B. Yaneva: Dokl. Bolgar. Akad. Nauk 26, 51 and 1449 (1973)
3.21 N.T. Kashukeev, G.A. Stanev: Dokl. Bolg. Akad. Nauk 26, 1445 (1973)
3.22 H. Maier-Leibnitz: Bayerische Akademie der Wissenschaften, Sitzungsberichte 1966, p. 173 (1967)
3.23 D. Bally, E. Tarina, N. Popa: Nucl. Instr. and Methods 127, 547 (1975)
3.24 D. Bally: Private communication (1976)
3.25 A. Steyerl: Z. Physik 254, 169 (1972)
3.26 A.R. Taylor: Preliminary Report on the Transmission of UCN along Stainless Steel Guide Tubes; Report (The University of Sussex 1975)
3.27 P. Ageron: Private communication (Institute Laue-Langevin, Grenoble 1976)
3.28 Yu.Yu. Kosvintsev, Yu.A. Kushnir, V.I. Morozov, A.P. Platonov: Preprint NIIAR P-268 (Dimitrovgrad 1976)
3.29 D. Winfield, J.M. Robson: Canad. J. Phys. 53, 667 (1975)
3.30 V.K. Ignatovich: JINR Communication R4-7055 (Dubna 1973)
3.31 I. Berceanu, V.K. Ignatovich: Vacuum 23, 441 (1973)
3.32 M. Brown, R. Golub, J.M. Pendlebury: Vacuum 25, 61 (1975)
3.33 B.N. Vinogradov, G.I. Terekhov: Transport of Ultracold Neutrons along Neutron Guides in the Diffusion Approximation; 3rd Conf. on Neutr. Phys. (Kiev 1975)
3.34 Yu.S. Zamyatin, A.G. Kolesov, E.N. Kulagin, V.I. Lushchikov, V.I. Morozov, V.N. Nefedov, Yu.N. Pokotilovsky, A.V. Strelkov, F.L. Shapiro: JINR communication R3-7946 (Dubna 1974)
3.35 A.V. Strelkov: Author's Abstract of Candidate's Dissertation, 3-5937 (JINR Dubna 1971)
3.36 A.V. Strelkov: 2nd Internat. School on Neutron Physics (Alushta 1974)
3.37 J.C. Bates, S. Roy: Private communication (1976)
3.38 A.V. Antonov, S.A. Antipov, A.I. Isakov, N.I. Ivanov, V.G. Kuznetsova, V.I. Mikerov, V.S. Sergeev, S.A. Startsev: Short Communications in Physics No. 11 (Lebedev Institute Moscow 1974) p. 11
3.39 A.V. Antonov, P.A. Belyaev, A.I. Isakov, A.A. Tikhomirov, S.A. Fridman, V.V. Shchaenko: Short Communications in Physics No. 12 (Lebedev Institute Moscow 1974) p. 30
3.40 A.V. Antonov, A.I. Isakov, V.N. Kovylnikov, V.I. Mikerov, S.A. Startsev, A. A. Tikhomirov: Short Communications in Physics No. 11 (Lebedev Institute Moscow 1974) p. 17
3.41 A.V. Antonov, A.I. Isakov, S.P. Kuznetsov, N.V. Linkova, V.I. Mikerov, A.D. Perekrestenko, S.A. Startsev: Short Communications in Physics No. 10 (Lebedev Institute Moscow 1974) p.14
3.42 J.C. Bates, S. Roy: Private communication (1972); S. Roy: Experimental and Computational Studies of Ultracold Neutrons; Ph.D. thesis (Manchester University, 1976)

3.43   N.T. Kashukeev: A Channel for Ultracold Neutrons at the Reactor IRT-2000, Sofia, with a Mechanical Moderator; Contribution ot the Symposium "Experience in the Experimentation and Use of Research Reactors"(Predeal, 1974)

4.1   L.D. Landau, E.M. Lifshits: Kvantavaya Mekhanika (Izd. Fizikomatemat. Lit. Moscow 1963); German Translation: Quantenmechanik (Akademie-Verlag, Berlin 1965) p. 571

4.2   K. Heinloth, T. Springer: The Measurement of the Total Cross Section of H_2O between -150 and +200^{0}C with Very Slow Neutrons; in Inelastic Scattering of Neutrons in Solids and Liquids (IAEA Vienna 1961) p. 323

4.3   K. Heinloth: Z. Physik 163, 218 (1961)

4.4   C.O. Fischer: Phys. Letters 30A, 393 (1969); Ber. Bunsenges. physik. Chem. 74, 696 (1970); Ber. Bunsenges. physik. Chem. 75, 361 (1971)

4.5   R. Lenk: phys. stat. sol. (b) 69, 369 (1975)

4.6   V.K. Ignatovich: JINR Communication R4-6681 (Dubna 1972)

4.7   V.K. Ignatovich: phys. stat. sol. (b) 71, 477 (1975)

4.8   A. Steyerl: Transmissionsmessungen mit ultrakalten Neutronen. Doctorate thesis (Technical University, Munich 1971)

4.9   S. Todireanu, V. Cioca: Rev. Roum. Phys. 15, 213 (1970)

4.10   G. Placzek: Phys. Rev. 93, 895 (1954)

4.11   K. Binder: phys. stat. sol. 41, 767 (1970)

4.12   K. Binder: Wirkungsquerschnitte für die Streuung von "ultrakalten" Neutronen. FRM Report PTHM-FRM-110 (Technical University, Munich 1970)

4.13   W. Mehringer: Z. Physik 210, 434 (1968)

4.14   K. Binder, H. Rauch: Nukleonik 11, 113 (1968)

4.15   P.H. Handel: Z. Naturforsch. 24a, 1646 (1969)

4.16   K. Binder: Z. Naturforsch. 26a, 432 (1971)

4.17   K. Binder: Z. Angewandte Phys. 32, 178 (1971)

4.18   T. Springer: Private communication (1976)

4.19   A. Heidemann: Z. Physik 238, 208 (1970); Z. Physik B 20, 385 (1975)

4.20   F. Mezei: Z. Physik 225, 146 (1972)

4.21   A. Steyerl: Neutronenstreuung an Blochwänden in polykristallinen Ferromagneten. FRM Report PDTUM-FRM-114 (Technical University, Munich 1974)

4.22   M. Lengsfeld: Transmissionsmessungen mit sehr langsamen Neutronen zur Bestimmung der Eigenschaften ferromagnetischer Domänen. Doctorate thesis (Technical University, Munich 1976)

4.23   T. Suzuki: Z. Angew. Physik 32, 75 (1971)

4.24   A. Steyerl: Z. Physik 252, 371 (1972)

4.25   A.V. Antonov, A.I. Isakov, V.I.Mikerov, S.A. Startsev: JETP Lett. 20, 289 (1974)

4.26   Yu.N. Pokotilovsky, Yu.V. Taran, F.L. Shapiro: JINR Preprint R3-9185 (1975)

5.1   I.I. Gurevich, P.E. Nemirovsky: Sov. Phys. -JETP 14, 838 (1962) (Zh. Eksper. i Teor. Fiz. 41, 1175 (1961))

5.2   L.L. Foldy: Bottles for Neutrons; in Preludes in Theor. Phys., ed. by A. de Shalit, H. Feshbach, L. van Hove (North Holland, Amsterdam 1966) p. 205

5.3   H. Solbrig: phys. stat. sol. (b) 51, 555 (1972)

5.4   V.K. Ignatovich, G.I. Terekhov: JINR Communication R4-9567 (Dubna 1976)

5.5   V.K. Ignatovich: JINR Communication R4-9007 (Dubna 1975)

5.6   L.V. Groshev, V.I. Lushchikov, S.A. Nikolayev, Yu.N. Panin, Yu.N. Pokotilovsky, A.V. Strelkov: JINR Communication R3-9534 (Dubna 1976)

5.7   Yu.N. Panin: Contribution to the 2nd International School on Neutron Physics (Alushta 1974)

5.8   M. Hetzelt: Private communication (1976)

5.9   V.A. Anikolenko, A.V. Antonov, A.I. Isakov, N.V. Lin'kova, A.D. Perekrestenko, V.E. Solodilov, S.A. Startsev, A.A. Tikhomirov: Short Communications in Physics No. 11 (Lebedev Institute, Moscow 1973) p. 40

5.10   A.V. Stepanov, A.V. Shelagin: Short Communications in Physics No. 1 (Lebedev Institute, Moscow 1974) p. 12

5.11   A.V. Stepanov: Teor. i Mat. Fiz. 22, 425 (1975)

5.12   V.K. Ignatovich, A.V. Stepanov: JINR Communication R4-7832 (Dubna 1974)

5.13 R. Lenk: phys. stat. sol. (a) $\underline{25}$, K141 (1974)
5.14 I.M. Frank: JINR Communication R4-8851 (Dubna 1975)
5.15 V.K. Ignatovich: Addendum to the Question of Ultracold Neutron Heating in Wall Collisions (to be published in phys. stat. sol.)
5.16 A.S. Gerasimov, V.K. Ignatovich, M.V. Kazarnovsky: JINR Preprint R4-6940 (Dubna 1973)
5.17 V.K. Ignatovich: JINR Communication R4-7831 (Dubna 1974)
5.18 J.J. Bikerman: Physical Surfaces (Academic Press, New York, London 1970)
5.19 I.M. Frank: JINR Preprint R3-7810 (Dubna 1974)
5.20 A.V. Stepanov: Short Communications in Physics No. 8 (Lebedev Institute, Moscow 1974) p. 34
5.21 A.V. Stepanov: Lebedev Inst. Nucl. Research Preprint P 0004 (Moscow 1975)
5.22 V.K. Ignatovich: JINR Preprint E4-8039 (Dubna 1974)
5.23 L. de Broglie: Non-Linear Wave Mechanics (Elsevier, Amsterdam, London 1960)
6.1   V.F. Ezhov, S.N. Ivanov, V.A. Knyaz'kov, V.M. Lobashov, V.A. Nazarenko, G.D. Porsev, A.P. Serebrov: Internat. Working Meeting on Diffraction of Polarized Neutrons (Swierk); In Proc. Zointe -IBJ 1974
6.2   B. Martin: Überlegungen und Vorversuche zur magnetischen Speicherung neutraler Teilchen. Doctorate thesis (University Bonn 1975)
6.3   E.M. Purcell, N.F. Ramsey: Phys. Rev. $\underline{78}$, 807 (1950)
6.4   R. Golub, J.M. Pendlebury: Contemp. Phys. $\underline{13}$, 519 (1972)
6.5   J.H. Christenson, J.W. Cronin, V.L. Fitch, R. Turlay: Phys. Rev. Lett. $\underline{13}$, 138 (1964)
6.6   N.F. Ramsey: Phys. Rev. $\underline{109}$, 225 (1958)
6.7   J.H. Smith: Ph.D. thesis (Harvard University 1951); J.H. Smith, E.M.Purcell, N.F. Ramsey: Phys. Rev. $\underline{108}$, 120 (1957)
6.8   P.D. Miller: 2nd International School on Neutron Physics (Alushta); In Proceedings JINR D3-7991 (Dubna 1974) p. 100
6.9   N.F. Ramsey: The electric and magnetic dipole moments of the neutron. Bull. Am. Phys. Soc., Ser. II, $\underline{21}$, 61 (1976)
6.10 T.D. Lee: Physics Reports $\underline{9}$, 143 (1974)
6.11 J. Frenkel, M.E. Ebel: Nucl. Phys. B83, 177 (1974)
6.12 R.N. Mohapatra, J.C. Pati: Phys. Rev. D11, 566 (1975)
6.13 N.F. Ramsey: Molecular Beams (University Press Oxford 1963)
6.14 A.I. Egorov, V.F. Ezhov, S.N. Ivanov, V.A. Knyaz'kov, V.M. Lobashov, V.A. Nazarenko, G.D. Porsev, A.P. Serebrov: Sov. J. Nucl. Phys. $\underline{21}$, 153 (1975) (Yad. Fiz. $\underline{21}$, 292 (1975))
6.15 Yu.V. Taran: JINR Communication R3-7147 (Dubna 1973)
6.16 Yu.V. Taran: JINR Communication R3-7149 (Dubna 1973)
6.17 Yu.V. Taran: JINR Depos. Communciation BI-3-7151 (Dubna 1973)
6.18 Yu.V. Taran: JINR Communication R3-7377 (Dubna 1973)
6.19 Yu.V. Taran: JINR Depos. Publ. BI-13-8441 (Dubna 1974)
6.20 Yu.V. Taran: JINR Communication R3-7785 (Dubna 1974)
6.21 Yu.V. Taran: JINR Communication R3-8442 (Dubna 1974)
6.22 Yu. V. Nikitenko, Yu.V. Taran: JINR Communication R3-7379 (Dubna 1973)
6.23 V.N. Efimov, V.K. Ignatovich: JINR Communication R4-8253 (Dubna 1974)
6.24 F.K. Smith, J.M. Pendlebury, R. Golub, J. Byrne, A. Steyerl, N.F. Ramsey, P.D. Miller, W.B. Dress, P. Perrin, P.G.H. Sanders: Search for the Neutron Electric Dipole Moment Using Bottled Neutrons. Research Proposal to the Inst. Laue-Langevin (Grenoble 1974)
6.25 I.I. Rabi, N.F. Ramsey, J. Schwinger: Rev. Mod. Phys. $\underline{26}$, 167 (1954)
6.26 Yu.V. Taran: JINR Communication R3-8577 (Dubna 1975)
6.27 A.R. Taylor: Ph.D. thesis (University of Sussex 1976)
6.28 Yu.V. Taran: JINR Communication R3-9307 (1975)
6.29 Yu.Yu. Kosvintsev, Yu.A. Kushnir, V.I. Morozov, G.I. Terekhov, Yu.N. Pokotilovsky: On the Possibility of the Application of Wall-type and Magnetic Bottles of Ultracold Neutrons for a Measurement of the Lifetime of the Free Neutron. Submitted to Prib. i Tekhn. Eksp.
6.30 I.M. Matora: Sov. J. Nucl. Phys. $\underline{16}$, 349 (1973) (Yad. Fiz. $\underline{16}$, 624 (1972)

6.31 V.K. Ignatovich: JINR Preprint E4-8404 (Dubna 1974)
6.32 B. Martin, W. Paul, U. Trinks, D. Cassel: Study of a Magnetic Storage Ring for Ultracold Neutrons. Research Proposal to the Inst. Laue-Langevin (Grenoble 1974)
6.33 K.-J. Kügler: Elektronen-Nachweis und Neutronen-Injektion am Neutronen-Speicherring-Experiment. Diploma thesis (Bonn University 1975)
6.34 Yu.G. Abov, V.F. Belkin, V.V. Vasil'ev, V.V. Vladimirsky, P.A. Krupchitsky, V.K. Rissukhin: Communication ITEF-44 (Inst. of Theor. and Experim. Phys. Moscow 1976)
6.35 B.G. Erozolimsky: Sov. Phys. - Usp. $\underline{18}$, 377 (1975) (Usp. Fiz. Nauk $\underline{116}$, 145 (1975))
6.36 A.N. Sosnovsky, P.E. Spivak, Yu.A. Prokofiev, I.E. Kutikov, Yu.P. Dobrinin: Nucl. Phys. $\underline{10}$, 395 (1959)
6.37 C.J. Christensen, A. Nielsen, A. Bahnsen, W.K. Brown, B.M. Rustad: Phys. Lett. $\underline{26B}$, 11 (1967)
6.38 C.G. Shull, K.W. Billman, F.A. Wedgwood: Phys. Rev. $\underline{153}$, 1415 (1967)
6.39 N.T. Kashukeev, N.B. Yaneva: Project 'KITKA'. Inst. of Nucl. Research and Nucl. Energy (Sofia 1975)
      (The survey of literature for this review was concluded in June, 1976.)

Classified Index

Springer Tracts in Modern Physics, Volumes 36-80

This cumulative index is based upon the Physics and Astronomy Classification Scheme (PACS) developed by the American Institute of Physics.

General

04 Relativity and Gravitation

Heintzmann, H., Mittelstaedt, P.: Physikalische Gesetze in beschleunigten Bezugssystemen (Vol. 47)
Stewart, J., Walker, M.: Black Holes: the Outside Story (Vol. 69)

05 Statistical Physics

Agarwal, G.S.: Quantum Statistical Theories of Spontaneous Emission and their Relation to Other Approaches (Vol. 70)
Graham, R.: Statistical Theory of Instabilities in Stationary Nonequilibrium Systems with Applications to Lasers and Nonlinear Optics (Vol. 66)
Haake, F.: Statistical Treatment of Open Systems by Generalized Master Equations (Vol. 66)

07 Specific Instrumentation

Godwin, R.P.: Synchrotron Radiation as a Light Source (Vol. 51)

The Physics of Elementary Particles and Fields

11 General Theory of Fields and Particles

Brandt, R.A.: Physics on the Light Cone (Vol. 57)
Dahmen, H.D.: Local Saturation of Commutator Matrix Elements (Vol. 62)
Ferrara, S., Gatto, R., Grillo, A.F.: Conformal Algebra in Space-Time and Operator Product Expansion (Vol. 67)
Jackiw, R.: Canonical Light-Cone Commutators and Their Applications (Vol. 62)
Kundt, W.: Canonical Quantization of Gauge Invariant Field Theories (Vol. 40)
Rühl, W.: Application of Harmonic Analysis to Inelastic Electron-Proton Scattering (Vol. 57)

Symanzik, K.: Small-Distance Behaviour in Field Theory (Vol. 57)
Zimmermann, W.: Problems in Vector Meson Theories (Vol. 50)

11.30 Symmetry and Conservation Laws

Barut, A.O.: Dynamical Groups and their Currents. A Model for Strong Interactions (Vol. 50)
Ekstein, H.: Rigorous Symmetrics of Elementary Particles (Vol. 37)
Gourdin, M.: Unitary Symmetry (Vol. 36)
Łopuszański, J.T.: Physical Symmetrics in the Framework of Quantum Field Theory (Vol. 52)
Pauli, W.: Continuous Groups in Quantum Machanics (Vol. 37)
Racah, G.: Group Theory and Spectroscopy (Vol. 37)
Rühl, W.: Application of Harmonic Analysis to Inelastic Electron-Proton Scattering (Vol. 57)
Wess, J.: Conformal Invariance and the Energy-Momentum Tensor (Vol. 60)
Wess, J.: Realisations of a Compact, Connected, Semisimple Lie Group (Vol. 50)

11.40 Currents and Their Properties

Furlan, G., Paver, N., Verzegnassi, C.: Low Energy Theorems and Photo- and Electroproduction Near Threshold by Current Algebra (Vol. 62)
Gatto, R.: Cabibbo Angle and SU_2 x SU_2 Breaking (Vol. 53)
Genz, H.: Local Properties of σ-Terms: A Review (Vol. 61)
Kleinert, H.: Baryon Current Solving SU (3) Charge-Current Algebra (Vol. 49)
Leutwyler, H.: Current Algebra and Lightlike Charges (Vol. 50)
Mendes, R.V., Ne'eman, Y.: Representations of the Local Current Algebra. A Constructional Approach (Vol. 60)
Müller, V.F.: Introduction to the Lagrangian Method (Vol. 50)
Pietschmann, H.: Introduction to the Method of Current Algebra (Vol. 50)
Pilkuhn, H.: Coupling Constants from PCAC (Vol. 55)
Pilkuhn, H.: S-Matrix Formulation of Current Algebra (Vol. 50)
Renner, B.: Current Algebra and Weak Interactions (Vol. 52)

Related Areas of Science and Technology

85.70 Magnetic Devices

Geophysics, Astronomy, and Astrophysics

95 Theoretical Astrophysics

97 Stars

Springer Tracts in Modern Physics

Ergebnisse der exakten Naturwissenschaften
Editor: G. Höhler

Volume 64
T. Springer, Jülich
Quasielastic Neutron Scattering for the Investigation of Diffusive Motions in Solids and Liquids

36 figures. II, 100 pages
1972
ISBN 3-540-05808-7

Contents: Scattering Theory.—Methodical and Experimental Aspects.—Monoatomic Liquids with Continuous Diffusion.—Jump Diffusion in Liquids.—Diffusion of Hydrogen in Metals.—Rotational Diffusion in Molecular Solids.— Molecular Liquids.— Polymeres and other Complicated Systems.—Effects of Coherent Scattering.—Quasielastic Scattering and other Methods.

Topics in Current Physics 3

Dynamics of Solids and Liquids by Neutron Scattering
Editors: S. Lovesey, Grenoble, T. Springer, Jülich

156 figures. Approx. 320 pages. 1977

The book contains seven chapters, written by internationally renowned experimental and theoretical scientists, on the motion of atoms and molecules in solids and liquids, studied by neutron spectroscopy. It includes structural phase transitions, phonons, magnetic excitations, collective excitations in monoatomic classical liquids, hydrogen diffusion in metals, rotations and tunneling in molecular crystals. The latter subjects are treated with special emphasis to problems related to physical chemistry. The inclusion of an introductory chapter and a uniform notation should facilitate the use of the book by scientists for whom neutron scattering is not a main research activity.

Springer-Verlag Berlin Heidelberg New York

Topics in Applied Physics

Springer-Verlag   Berlin   Heidelberg   New York